数据结构实践训练教程

主编 刘光然
编著 徐棣 罗梅 李典蔚

南开大学出版社
天　津

图书在版编目(CIP)数据

数据续构实践训练教程 / 刘光然主编. —天津：南开大学出版社，2009.4(2017.8重印)
ISBN 978-7-310-03113-9

Ⅰ.数… Ⅱ.刘… Ⅲ.数据结构—教材 Ⅳ.TP311.12

中国版本图书馆 CIP 数据核字(2009)第 035431 号

版权所有　侵权必究

南开大学出版社出版发行
出版人：刘立松
地址：天津市南开区卫津路 94 号　邮政编码：300071
营销部电话：(022)23508339　23500755
营销部传真：(022)23508542　邮购部电话：(022)23502200

＊
天津午阳印刷有限公司印刷
全国各地新华书店经销
＊
2009 年 4 月第 1 版　2017 年 8 月第 2 次印刷
787×1092 毫米　16 开本　17.75 印张　455 千字
定价：38.00 元(含光盘一张)

如遇图书印装质量问题，请与本社营销部联系调换，电话：(022)23507125

内容提要

本书深入浅出地阐述数据结构的基础知识，并根据每章的知识点，精选出具有针对性、实用性、普及性的经典实训项目，让学生在训练过程中边学边练，在不知不觉中得以全面提高计算机素质。最后还给出了3个用以训练学生综合运用能力的综合项目案例。

本书的特点是"以能力培养为核心，以技能训练为主线，以实践项目为载体，以理论知识为支撑"；注重理论和实践相结合，用理论指导实践，在实践中理解并运用理论；保证各个实训项目的科学性、实践性、实用性和可操作性，附赠一张CD-ROM配套程序光盘。

本书既可作为高等学校应用型本科和高等职业院校计算机相关专业学生的实训教材，也可作为计算机程序爱好者的自学参考书。

前　言

在大力发展职业教育的今天，各高校对学生实际技能的培养越来越重视，尤其是应用型本科学校和高等职业院校更是把培养学生的动手能力和应有的职业技能作为首要职责，于是教学改革不断深入。随着 CDIO 教学模式的引进和推广，越来越多的学者开始进行研究和尝试，基于项目的教学培养方式得到普遍认可并获得广泛的应用，基于项目的实践教学已经成为目前各高校赖以培养学生实际操作技能的主要法宝之一。基于此，结合多年实际教学经验，我们尝试着编写了本教程。

在本教程的编写过程中，我们遵循"以能力培养为核心，以技能训练为主线，以实践项目为载体，以理论知识为支撑"的思想；注重理论和实践相结合，用理论指导实践，在实践中理解并运用理论，保证各个实训项目的科学性、实践性、实用性和可操作性。

本教程共计 8 章。第 1 章介绍线性表的概念，提供了学生成绩管理系统、考试报名管理系统和约瑟夫生者死者游戏等 4 个项目案例；第 2 章介绍栈和队列的概念，提供了勇闯迷宫游戏、N 皇后问题和停车场管理系统等 3 个项目案例；第 3 章介绍串的概念，给出了关键字检索和四元线性方程组求解等 2 个实训案例；第 4 章介绍树和二叉树的概念，列举了家谱管理、表达式求值、图像压缩编码优化等 3 个实际问题及其解决方案；第 5 章介绍了图的基本知识，给出了公交路线管理模拟、导航最短路径查询、电网建设造价模拟、软件工程进度规划等日常实用案例；第 6 章介绍了查找的基本概念，给出了顺序查找、折半查找、二叉排序树和哈希查找等 4 个简单案例；第 7 章介绍了排序的有关概念，并将相关的排序算法融为一个大的程序，即各种排序算法的比较程序；第 8 章是综合应用篇，我们给出了迷宫益智游戏、景区旅游信息管理系统（Console 版本）、景区旅游信息管理系统（MFC 版本）等 3 个要求利用所学知识完成的综合性实训案例，以提高学生的综合运用所学知识的实际能力。

在学习本教程时，需要读者具有一定的 C++编程基础和阅读程序代码的能力。

本教程由刘光然总体策划；各章节基本概念的内容由徐棣和罗梅确定，由罗梅老师主笔；实训案例由徐棣和李典蔚设计，程序源代码由李典蔚老师编写和调试（详见本书配套的 CD-ROM 光盘）；全书最后由刘光然、徐棣老师统一修改定稿。

本教程在编写过程中，作者参阅了相关的书籍和网络资源，在此向所有这些书籍、资料的作者们表示衷心的感谢，同时感谢我的朋友和同事以及张燕老师、尹建国老师对本教程的写作和出版提供的帮助。

读者在学习过程中如需与作者联系，请发邮件至 Liuguangran@163.com。由于作者的能力有限，水平不高，错误之处在所难免，恳请读者朋友批评指正。

刘光然
2009 年元旦于天津

目 录

第一章 线性表

- 1.1 实践目的和要求 ... 1
  - 1.1.1 实践目的 ... 1
  - 1.1.2 实践要求 ... 1
- 1.2 基本概念 ... 1
  - 1.2.1 线性表的定义 ... 1
  - 1.2.2 线性表的顺序存储结构 ... 2
  - 1.2.3 线性表的链式存储结构 ... 2
  - 1.2.4 线性表的基本运算 ... 4
- 1.3 实践案例 ... 5
  - 1.3.1 学生成绩管理系统 ... 5
  - 1.3.2 考试报名管理系统 ... 15
  - 1.3.3 约瑟夫生者死者游戏 ... 26
  - 1.3.4 约瑟夫双向生死游戏 ... 31
- 1.4 巩固提高 ... 38

第二章 栈和队列

- 2.1 实践目的和要求 ... 39
  - 2.1.1 实践目的 ... 39
  - 2.1.2 实践要求 ... 39
- 2.2 基本概念 ... 40
  - 2.2.1 栈 ... 40
  - 2.2.2 队列 ... 41
- 2.3 实践案例 ... 44
  - 2.3.1 勇闯迷宫游戏 ... 44
  - 2.3.2 N 皇后问题 ... 51
  - 2.3.3 停车场管理系统 ... 56
- 2.4 巩固提高 ... 67

第三章 串

- 3.1 实践目的和要求 ... 68
  - 3.1.1 实践目的 ... 68
  - 3.1.2 实践要求 ... 68
- 3.2 基本概念 ... 68
  - 3.2.1 串的定义 ... 68

3.2.2 串的存储结构 ··· 69
　　　3.2.3 串的基本运算 ··· 70
　3.3 实践案例 ·· 71
　　　3.3.1 关键字检索系统 ·· 71
　　　3.3.2 四元线性方程组求解 ··· 75
　3.4 巩固提高 ·· 79

第四章　树和二叉树

　4.1 实践目的和要求 ·· 80
　　　4.1.1 实践目的 ·· 80
　　　4.1.2 实践要求 ·· 80
　4.2 基本概念 ·· 81
　　　4.2.1 树 ·· 81
　　　4.2.2 二叉树 ·· 84
　4.3 实践案例 ·· 87
　　　4.3.1 家谱管理系统 ··· 87
　　　4.3.2 表达式求值问题 ·· 95
　　　4.3.3 图像压缩编码优化问题 ··· 99
　4.4 巩固提高 ·· 107

第五章　图

　5.1 实践目的和要求 ·· 108
　　　5.1.1 实践目的 ·· 108
　　　5.1.2 实践要求 ·· 108
　5.2 基本概念 ·· 108
　　　5.2.1 图的定义 ·· 108
　　　5.2.2 图的相关术语 ·· 109
　　　5.2.3 图的存储结构 ·· 110
　　　5.2.4 图的遍历 ·· 111
　　　5.2.5 图的基本运算 ·· 113
　5.3 实践案例 ·· 113
　　　5.3.1 公交路线管理模拟系统 ·· 113
　　　5.3.2 最短路径导航查询系统 ·· 127
　　　5.3.3 电网建设造价模拟系统 ·· 134
　　　5.3.4 软件工程进度规划系统 ·· 144
　5.4 巩固提高 ·· 154

第六章　查找

　6.1 实践目的和要求 ·· 155
　　　6.1.1 实践目的 ·· 155
　　　6.1.2 实践要求 ·· 155
　6.2 基本概念 ·· 156
　　　6.2.1 查找的概念 ··· 156
　　　6.2.2 线性表的查找 ·· 156
　　　6.2.3 树表的查找 ··· 157
　　　6.2.4 哈希表的查找 ·· 159

6.3 实践案例 ············ 161
6.3.1 顺序查找 ············ 161
6.3.2 折半查找 ············ 163
6.3.3 二叉排序树 ············ 166
6.3.4 哈希查找 ············ 172
6.4 巩固提高 ············ 176

第七章 排序
7.1 实践目的和要求 ············ 178
7.1.1 实践目的 ············ 178
7.1.2 实践要求 ············ 178
7.2 基本概念 ············ 178
7.2.1 排序的概念 ············ 178
7.2.2 插入排序 ············ 179
7.2.3 选择排序 ············ 180
7.2.4 交换排序 ············ 181
7.2.5 归并排序 ············ 182
7.2.6 基数排序 ············ 183
7.2.7 各种排序方法比较 ············ 184
7.3 实践案例 ············ 184
7.3.1 系统简介（8种排序算法比较案例） ············ 184
7.3.2 设计思路 ············ 185
7.3.3 程序清单 ············ 185
7.3.4 运行结果 ············ 197
7.4 巩固提高 ············ 197

第八章 综合篇
8.1 目的和要求 ············ 199
8.1.1 实践目的 ············ 199
8.1.2 实践要求 ············ 199
8.2 相关概念 ············ 199
8.3 实践案例 ············ 200
8.3.1 迷宫益智游戏 ············ 200
8.3.2 景区旅游信息管理系统（Console 版本） ············ 232
8.3.3 景区旅游信息管理系统（MFC 版本） ············ 252

参考文献 ············ 272

第一章 线性表

线性表是数据结构中最简单、最常用的一种线性结构，也是学习数据结构全部内容的基础，其掌握的好坏直接影响着后继知识的学习。本章通过四个模拟实例来学习线性表的顺序和链式存储结构，首先通过使用有关数组的操作实现学生成绩管理，其次通过使用有关线性链表的操作实现考试报名管理，然后通过使用循环链表的操作实现约瑟夫生者死者游戏。

1.1 实践目的和要求

1.1.1 实践目的

1）掌握线性表顺序存储结构和链式存储结构的思想和特点；
2）掌握上机调试线性表的基本方法；
3）掌握线性表的插入、删除、查找以及线性表合并等运算在顺序存储结构和链式存储结构上的实现。

1.1.2 实践要求

1）认真阅读和掌握本章的程序，注意插入、删除时元素的移动原因、方向及先后顺序，理解不同的函数形参与实参的传递关系。
2）上机运行本章的程序，保存和打印出程序的运行结果，并结合程序进行分析。
3）按照你对线性表的操作需要，重新改写程序并运行。
4）重点理解链式存储的特点及指针的含义。
5）注意比较顺序存储与链式（单向、双向、循环链表）存储的特点及实现方法。
6）注意比较带头结点、无头结点链表实现插入、删除运算时的区别。
7）单向链表的操作是数据结构的基础，要注意对这部分的常见算法的理解。

1.2 基本概念

1.2.1 线性表的定义

线性表是具有相同数据类型的数据元素的一个有限序列。该序列中所含元素的个数叫作线性表的长度，用 n（n≥0）表示。当 n=0 时，表示线性表是一个空表，即表中不含任何元素。设序列中第 i（i 表示位序）个元素为 e_i（1≤i≤n），则线性表的一般表示为：（e_1, e_2, …, e_{i-1}, e_i, e_{i+1}, …, e_n），其中：e_1 为第一个元素，又称表头元素，e_2 为第二个元素，e_n 为最后一个元素，又称表尾元素。对于非空线性表，除第一个元素之外，表中的每个元素均有且只有一个直接前驱；除最后一个元素之外，表中每个元素均有且只有一个直接后继。

一个线性表可以用一个标识符来命名,如用 L 来命名上面的线性表,则有:
$$L=(e_1, e_2, \cdots, e_{i-1}, e_i, e_{i+1}, \cdots, e_n)$$

线性表最重要的性质是线性表中的元素在位置上是有序的,即第 i 个元素 e_i 处在第 i-1 个元素 e_{i-1} 的后面和第 i+1 个元素 e_{i+1} 的前面,这种位置上的有序性就是一种线性关系,所以线性表是一个线性结构,用二元组表示为:

L=(D, R)

D={e_i|1≤i≤n, n≥0, e_i 属于 ElemType 类型}   //ElemType 是 C/C++的类型标识符

R={r}, r={<e_i, e_{i+1}>|1≤i≤n-1}

当一个线性表的元素升序或降序排列时,称为有序线性表,简称为有序表。

有多种存储方式能将线性表存储在计算机内,其中最常用的是顺序存储结构和链式存储结构,分别简称为顺序表和链表。

1.2.2 线性表的顺序存储结构

顺序存储结构是最简单的存储方式,把线性表中的所有元素按照其逻辑顺序依次存储到计算机存储器中的一块连续的存储空间中。通常使用一个足够大的数组,从数组的第一个元素开始,将线性表的结点依次存储在数组中。假设线性表的元素类型为 ElemType,线性表长度为 n,则整个线性表所占用存储空间的大小为 n* sizeof(ElemType)。

在定义一个线性表的顺序存储结构时,需要定义数组来存储线性表中的所有元素和定义一个整型变量来存储线性表的长度。

○顺序存储结构的特点

逻辑关系上相邻的两个元素在内存中的物理位置上也相邻。

○顺序存储结构的优点

①无需为表示结点间的逻辑关系而额外增加存储空间;

②能随机访问线性表中的任一元素。线性表的第 i 个元素 e_i 的存储地址可以通过以下公式求得:LOC(e_i)=LOC(e_1)+(i-1)*sizeof(ElemType),其中 LOC(e_1)是线性表的表头元素 e_1 的存储位置,通常称作线性表的起始位置或基地址。

○顺序存储结构的缺点

①由于要求占用连续的存储空间,存储分配只能预先进行,线性表的最大长度需预先确定,从而浪费大量的存储空间;

②插入与删除运算的效率很低。为了保持线性表中的元素的顺序,执行线性表的插入和删除操作时需要移动其他大量元素,这对于插入和删除操作频繁的线性表及每个元素所占字节较大的问题将导致系统的运行速度难以提高;

③线性表的顺序存储结构的存储空间不便于扩充。当一个线性表按顺序存储结构存储后,如果在原线性表的存储空间后找不到与之连续的可用空间,但还需要插入新的元素,则会发生"上溢"错误。

1.2.3 线性表的链式存储结构

链式存储结构即用一组任意的存储单元存储线性表中的元素,每个存储结点不仅包含所存储元素本身的信息即数据域,而且包含元素之间的逻辑关系的信息(前驱结点包含后继结

点的地址信息）即指针域。一般地，每个存储结点有一个或多个这样的指针域。若一个结点中的某个指针域不需要指向任何结点，则将它的值置为空指针，用常量 NULL 表示。

根据指针域的不同和结点构造指针域方法的不同，链式存储结构有如下三种形式（每种形式均分为带头结点结构和不带头结点结构两种结构）。

1.2.3.1 单链表（线性链表）

链表的每个存储结点除要存储线性表元素本身的信息外，只设置一个指针域用来指向其直接后继结点。

如图 1.1 所示为带头结点的单链表结构。

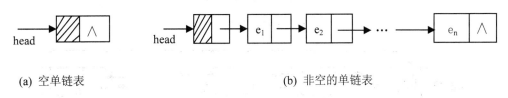

(a) 空单链表　　　　　　　　　　(b) 非空的单链表

图1.1　单链表的存储结构

〇 带头结点的单链表的特点

每个单链表都有一个头指针，整个链表的存取必须从头指针开始，头指针指向头结点的位置，头结点的指针域指向第一个元素的位置，最后的结点指针为空。当链表为空时，头结点的指针为空值；链表非空时，头结点指向第一个结点。

1.2.3.2 循环链表

循环链表是另一种形式的链式存储结构，是单链表的变形。它的特点是表中最后一个结点的指针域指向头结点，整个链表形成一个环。因此，从表中任意一个结点出发都可以找到表中的其他结点。

如图 1.2 所示为带头结点的循环链表结构。

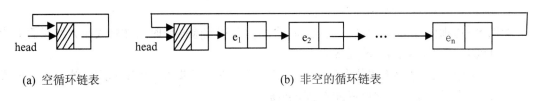

(a) 空循环链表　　　　　　　　　　(b) 非空的循环链表

图1.2　循环链表的存储结构

循环链表和单向链表基本相同，差别仅在于算法中判断链表是否已到达末结点的条件不是结点的指针是否为空，而是它的指针是否等于头指针，即循环单向链表末结点的指针不为空，而是指向了表的前端。

〇 循环链表的特点

只要知道表中某一结点的地址，就可搜寻到所有其他结点的地址。

1.2.3.3 双向链表

双向链表是线性表的另一种形式的链式存储结构,双向链表的存储结点中除要存储线性表元素本身的信息外,还设置有两个指针域,分别指向其直接后继结点和直接前驱结点。双向链表克服了单链表的单向性的缺点。

如图 1.3 所示为带头结点的双向链表结构。

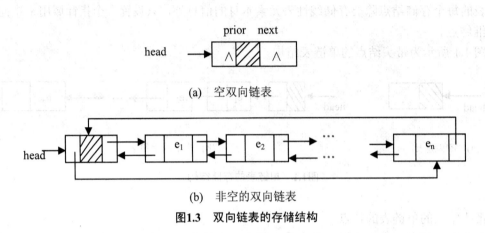

图1.3 双向链表的存储结构

双向链表也有循环链表形式,即链表中存在两个环。一个结点的直接前驱的直接后继和该结点的直接后继的直接前驱都是指向该结点的,即假定某存储结点的指针域为 p,则有:

$$p == p\text{->}prior\text{->}next == p\text{->}next\text{->}prior$$

○链式存储结构的特点

逻辑关系上相邻的两个元素在内存中的物理位置通过指针来表示。

○链式存储结构的优点

①由于不要求占用连续的存储空间,不需预先确定线性表的最大长度,存储分配可以动态进行,不会造成空间的浪费;

②插入与删除运算的效率高。执行线性表的插入和删除操作时只需修改相关结点的指针域即可,不需移动其他元素;

③线性表的链式存储结构的存储空间便于扩充。由于不要求存储空间的连续,因此只要内存空间还有剩余就可以插入新的元素。

○链式存储结构的缺点

①由于要存储相邻结点的地址信息,所以存储空间的开销较大;

②不具有顺序存储结构的随机存取特点,访问任何元素均需从头结点开始,影响访问效率。

1.2.4 线性表的基本运算

线性表是一个相当灵活的数据结构,它的长度可以根据需要增长或缩短。设基本线性表 L=(e_1, e_2, …, e_{i-1}, e_i, e_{i+1}, …, e_n),对基本线性表 L 的基本运算如下:

1) InitList(&L) 初始化:构造一个新的空线性表 L。

2）DestroyList（&L）撤销：销毁一个已有线性表 L。

3）ClearList（&L）清空：将一个已有线性表 L 置为空表。

4）ListEmpty（L）判空：判断一个已有线性表 L 是否为空表，是空表则返回 TRUE，否则返回 FALSE。

5）ListLength（L）求表长：返回一个已有线性表 L 中元素的个数。

6）GetElem（L，i，&e）取元素：返回一个已有线性表 L 中的第 i（1≤i≤ListLength（L））个元素的值。

7）LocateElem（L，e，compare（））元素定位：返回 L 中第一个与 e 满足关系 compare（）的数据元素的位置，若不存在这样的数据元素则返回 0。

8）PriorElem（L，cur_e，&pre_e）取直接前驱：若线性表 L 中存在元素 cur_e，且不是表头元素，则用 pre_e 返回它的直接前驱，否则操作失败。

9）NextElem（L，cur_e，&next_e）取直接后继：若线性表 L 中存在元素 cur_e，且不是表尾元素，则用 next_e 返回它的直接后继，否则操作失败。

10）ListInsert（&L，i，e）插入元素：在线性表 L 的第 i（1≤i≤ListLength（L）+1）个元素之前插入数据元素 e。线性表 L 长度加 1。

11）ListDelete（&L，i，&e）删除元素：删除线性表 L 中的第 i（1≤i≤ListLength（L））个元素。线性表 L 的长度减 1。

12）ListTraverse（L，visit（））遍历：对给定线性表 L 中的每一个数据元素依次调用 visit（）进行遍历。

13）CopyList（L，C）复制：将给定线性表 L 复制为线性表 C。

14）Merge（A，B，C）合并：将给定的线性表 A 和 B 合并为线性表 C。

根据存储方式的不同，上述运算的实现方法也不一样。

1.3 实践案例

1.3.1 学生成绩管理系统

1.3.1.1 系统简介

学生成绩管理是学校教务部门日常工作的重要组成部分，其处理信息量很大。本项目是对学生成绩管理的简单模拟，用控制台选项的选择方式完成下列功能：输入学生数据；输出学生数据；学生数据查询；添加学生数据；修改学生数据；删除学生数据。

1.3.1.2 设计思路

本项目的实质是完成对学生成绩信息的建立、查找、插入、修改、删除等功能。项目在设计时应首先确定系统的数据结构，定义类的成员变量和成员函数；然后实现各成员函数以完成对数据操作的相应功能；最后完成主函数以验证各个成员函数的功能并得出运行结果。

○ **数据结构**

分析发现，本项目的数据是一组学生的成绩信息，每条学生的成绩信息可由学号、姓名和成绩组成。而且这组学生的成绩信息具有相同特性，属于同一数据对象，相邻数据元素之间存在序偶关系。由此可以看出，这些数据具有线性表中数据元素的性质，所以该系统的数据适合采用线性表来存储。

线性表有两种存储方式，顺序表是线性表的顺序存储结构，是指用一组连续的内存单元依次存放线性表的数据元素。在顺序存储结构下，逻辑关系相邻的两个元素在物理位置上也相邻，这是顺序表的特点。顺序表适宜于作查找这样的静态操作，其优点是存储密度大，存储空间利用率高，缺点是插入或删除元素时不方便。

本项目确定使用线性表的顺序存储结构。

由于 C 语言的数组类型也有随机存储的特点，一维数组的机内表示就是顺序结构。因此，本项目可用 C 语言的一维数组实现线性表的顺序存储。

本系统的数据结构决定了其使用时的特性，根据顺序存储的特点可以发现，本系统可以方便的随机存取表中任一元素 O(1)，存储空间使用紧凑；缺点是在插入、删除某一元素时，需要移动大量元素 O(n)，预先分配空间需按最大空间分配，利用不充分，数组容量难以扩充。

○**程序设计和头文件定义**

经过上述分析，可确定系统的数据结构为数组，数组中的元素是学生成绩信息类。Student 类的成员变量和成员函数在头文件 Student.h 中定义，具体描述为：

```
class Student
{
public:
    void      Creat     ( Student stu[] );
    void      Insert    ( Student stu[] );
    void      Delete    ( Student stu[] );
    void      Lookup    ( Student stu[] );
    void      Update    ( Student stu[] );
    void      Stat      ( Student stu[] );
    int       Length    ( Student stu[] );
    void      Print     ( Student stu[] );

    Student();

private:
    string    name;              //姓名
    long      num;               //学号
    float     score;             //成绩
};
```

1.3.1.3 程序清单

```
/***************************************************************
    包含输入输出头文件 iostream 和字符串头文件 string，下面的案例将不再重复包含
****************************************************************/
```

```cpp
#include "iostream"
#include "string"
using namespace std;

#include "Student.h"

/***********************************************************************
    Student 类的构造函数，初始化元素数据
***********************************************************************/
Student::Student()
{
    name = '?';
    num = 0;
    score = 0;
}


/***********************************************************************
    创建学生成绩信息的线性表（数组）
***********************************************************************/
void Student::Creat(Student stu[])
{
    cout << "请输入学生人数：";
    int n;
    cin >> n;
    cout<<"姓名"<<'\t'<<"学号"<<'\t'<<"成绩"<<endl;
    for(int i = 1; i <= n; i ++)
    {
        string newname;
        long newnum;
        float newscore;
        cin >> newname;
        stu[i].name = newname;
        cin >> newnum;
        stu[i].num = newnum;
        cin >> newscore;
        stu[i].score = newscore;
    }
}
```

```
/************************************************************************
    输出线性表中所有的学生成绩信息
************************************************************************/
void Student::Print(Student stu[])
{
    cout << "姓名" << '\t' << "学号" << '\t' << "成绩" << '\n';
    int i = 1;
    while (stu[i].num) {
        cout << stu[i].name << '\t' << stu[i].num << '\t' << stu[i].score << '\n';
        i++;
    }
    cout << '\n';
}


/************************************************************************
    在线性表的某个位置上插入学生信息
************************************************************************/
void Student::Insert(Student stu[])
{
    if (Student::Length(stu) == 100) {
        cout << "存储空间已满，不能进行插入操作！" << '\n';
    }
    else
    {
        cout << "请输入要插入的位置：" ;
        int m;
        cin >> m;
        int n = Student::Length(stu);
        if (m > n + 1) {
            cout << "插入位置不正确，请重新输入！" << '\n';
            Student::Insert(stu);
        }
        else {
            for(int i = n; i >= m; i --)                //数组中的数据依次后移
            {
                stu[i + 1].name = stu[i].name;
                stu[i + 1].num = stu[i].num;
                stu[i + 1].score = stu[i].score;
            }
            cout << "请依次输入姓名，学号，成绩" << '\n';
```

```cpp
            string newname;
            long newnum;
            float newscore;
            cin >> newname;
            stu[m].name = newname;
            cin >> newnum;
            stu[m].num = newnum;
            cin >> newscore;
            stu[m].score = newscore;
        }
    }
}

/************************************************************************
    根据学号删除线性表中对应的学生信息
************************************************************************/
void Student::Delete(Student stu[])
{
    cout << "请输入你要删除的学号：";
    long num;
    cin >> num;
    int i = 1;
    if (!stu[i].num) {
        cout << "你要删除的学号不存在，请重新输入！" << '\n';
        Student::Delete(stu);
    }
    while (stu[i].num) {
        if (stu[i].num == num) {
            int n = Student::Length(stu);
            for(int j = i; j < n; j ++)                //数组中数据依次前移
            {
                stu[i].name = stu[i + 1].name;
                stu[i].num = stu[i + 1].num;
                stu[i].score = stu[i + 1].score;
            }
            stu[j].name = '?';
            stu[j].num = 0;
            stu[j].score = 0;
            break;
        }
```

```cpp
        i ++;
    }
}

/************************************************************************
    根据学号查询线性表中某个学生成绩信息
************************************************************************/
void Student::Lookup(Student stu[])
{
    cout << "请输入你要查询的学号：";
    long num;
    cin >> num;
    int i = 1;
    while (stu[i].num) {
        if (stu[i].num == num) {
            cout << "姓名" << '\t' << "学号" << '\t' << "成绩" << '\n';
            cout << stu[i].name << '\t' << stu[i].num << '\t' << stu[i].score << '\n';
            break;
        }
        i ++;
    }
    if (!stu[i].num) {
        cout << "你要查询的学号不存在，请重新输入！" << '\n';
        Student::Lookup(stu);
    }
    cout << '\n';
}

/************************************************************************
    修改线性表中某个位置的学生信息
************************************************************************/
void Student::Update(Student stu[])
{
    cout << "请输入你要修改的位置：";
    int m;
    cin >> m;
    if (m > Student::Length(stu)) {
        cout << "你要修改的位置不存在，请重新输入！" << '\n';
        Student::Update(stu);
    }
```

```cpp
        else {
            cout << "请依次输入更改后的姓名，学号，成绩" << '\n';
            string newname;
            long newnum;
            float newscore;
            cin >> newname;
            stu[m].name = newname;
            cin >> newnum;
            stu[m].num = newnum;
            cin >> newscore;
            stu[m].score = newscore;
        }
}

/************************************************************************
    查询线性表中有效数据的长度
************************************************************************/
int Student::Length(Student stu[])
{
    int i = 1;
    while (stu[i].num) {
        i ++;
    }
    return (i - 1);
}

/************************************************************************
    统计各成绩段的学生信息
************************************************************************/
void Student::Stat(Student stu[])
{
    cout << "不及格的学生是： " << '\n';
    cout << "姓名" << '\t' << "学号" << '\t' << "成绩" << '\n';
    int i = 1;
    while (stu[i].num) {
        if (stu[i].score < 60) {
            cout << stu[i].name << '\t' << stu[i].num << '\t' << stu[i].score << '\n';
        }
        i ++;
```

```cpp
        }
        cout << '\n';

        cout << "成绩为"良"的学生是:" << '\n';
        cout << "姓名" << '\t' << "学号" << '\t' << "成绩" << '\n';
        i = 1;
        while (stu[i].num) {
            if (stu[i].score >= 60 && stu[i].score < 75) {
                cout << stu[i].name << '\t' << stu[i].num << '\t' << stu[i].score << '\n';
            }
            i ++;
        }
        cout << '\n';

        cout << "成绩为"中"的学生是:" << '\n';
        cout << "姓名" << '\t' << "学号" << '\t' << "成绩" << '\n';
        i = 1;
        while (stu[i].num) {
            if (stu[i].score >= 75 && stu[i].score < 90) {
                cout << stu[i].name << '\t' << stu[i].num << '\t' << stu[i].score << '\n';
            }
            i ++;
        }
        cout << '\n';

        cout << "成绩为"优"的学生是:" << '\n';
        cout << "姓名" << '\t' << "学号" << '\t' << "成绩" << '\n';
        i = 1;
        while (stu[i].num) {
            if (stu[i].score >= 90) {
                cout << stu[i].name << '\t' << stu[i].num << '\t' << stu[i].score << '\n';
            }
            i ++;
        }
        cout << '\n';
}

/***********************************************************************
    主函数,调用类 Student 的成员函数以实现相应功能
***********************************************************************/
```

```cpp
void main()
{
    cout << "首先建立学生管理系统！" << '\n';
    Student stu[101];
    stu[0].Creat(stu);
    stu[0].Print(stu);

    int j=100;
    cout << "请选择您要进行的操作（1 为插入，2 为删除，3 为查找，4 为修改，5 为统计，0 为取消操作）";
    while(j)
    {
        cout << "请选择您要进行的操作：";
        cin >> j;
        switch(j){
        case 1:
            {
                stu[0].Insert(stu);
                stu[0].Print(stu);
                break;
            }
        case 2:
            {
                stu[0].Delete(stu);
                stu[0].Print(stu);
                break;
            }
        case 3:
            {
                stu[0].Lookup(stu);
                break;
            }
        case 4:
            {
                stu[0].Update(stu);
                stu[0].Print(stu);
                break;
            }
        case 5:
            {
```

```
                    stu[0].Stat(stu);
                    break;
            }
        default:
            break;
        }
    }
    cout <<"线性表中共有 "<<stu[0].Length(stu)<<" 个学生";
}
```

1.3.1.4 运行结果

```
首先建立学生管理系统!
请输入学生人数：3
姓名    学号    成绩
stu1 1 90
stu2 2 60
stu3 3 55
姓名      学号     成绩
stu1       1        90
stu2       2        60
stu3       3        55

请选择您要进行的操作(1为插入，2为删除，3为查找，4为修改，5为统计，0为取消操作)
请选择您要进行的操作：1
请输入要插入的位置：3
请依次输入姓名，学号，成绩
stu4 4 80
姓名      学号     成绩
stu1       1        90
stu2       2        60
stu4       4        80
stu3       3        55

请选择您要进行的操作：2
请输入你要删除的学号：3
姓名      学号     成绩
stu1       1        90
stu2       2        60
stu4       4        80

请选择您要进行的操作：3
请输入你要查询的学号：3
你要查询的学号不存在，请重新输入!
请输入你要查询的学号：4
姓名      学号     成绩
stu4       4        80
```

```
请选择您要进行的操作:4
请输入你要修改的位置:3
请依次输入更改后的姓名,学号,成绩
stu4 3 80
姓名    学号    成绩
stu1    1       90
stu2    2       60
stu4    3       80

请选择您要进行的操作:5
不及格的学生是:
姓名    学号    成绩

成绩为"良"的学生是:
姓名    学号    成绩
stu2    2       60

成绩为"中"的学生是:
姓名    学号    成绩
stu4    3       80

成绩为"优"的学生是:
姓名    学号    成绩
stu1    1       90

请选择您要进行的操作:0
线性表中共有 3 个学生
```

1.3.2 考试报名管理系统

1.3.2.1 系统简介

考试报名工作给各高校报名工作带来了新的挑战,给教务管理部门增加了很大的工作量。本项目是对考试报名管理的简单模拟,用控制台选项的选择方式完成下列功能:输入考生信息;输出考生信息;查询考生信息;添加考生信息;修改考生信息;删除考生信息。

1.3.2.2 设计思路

本项目的实质是完成对考生信息的建立、查找、插入、修改、删除等功能。项目在设计时应首先确定系统的数据结构,定义类的成员变量和成员函数;然后实现各成员函数以完成对数据操作的相应功能;最后完成主函数以验证各个成员函数的功能并得出运行结果。

○**数据结构**

分析发现,本项目的数据是一组考生信息,每条考生信息可由准考证号、姓名、性别、年龄、报考类别等信息组成。而且这组学生的成绩信息具有相同特性,属于同一数据对象,相邻数据元素之间存在序偶关系。由此可以看出,这些数据具有线性表中数据元素的性质,所以该系统的数据适合采用线性表来存储。

链式存储是线性表的另一种表示方法,由于它不要求逻辑上相邻的元素在物理位置上也相邻,因此其优点是插入或删除元素时很方便,使用灵活。缺点是存储密度小,存储空间利用率低。事实上,链表插入、删除运算的快捷是以空间代价来换取时间。

若线性表的长度变化不大,且其主要操作是查找,则采用顺序表;若线性表的长度变化较大,且其主要操作是插入、删除操作,则采用链表。本项目对考生数据主要进行插入、删除、修改等操作，所以采用链式存储结构比较适合。

本项目确定使用线性表链式存储结构。

系统的数据结构决定了其使用时的特性，根据链式存储的特点可以发现，本系统可以方便的进行插入、删除、修改等操作；缺点是失去了顺序表可随机存取的特点。

○**程序设计和头文件定义**

经过上述分析，可确定系统的数据结构为链表，链表的结点作为一个类用以存储考生信息。具体的成员变量和成员函数在头文件 Examinee.h 中定义为：

```
/******************************************************************
    Node 类作为链表的结点，用以存储单个考生信息
******************************************************************/
class Node
{
    friend class Examinee;
public:
    long    examnum;            //准考证号
    string  name;               //姓名
    string  sex;                //性别
    int     age;                //年龄
    string  type;               //报考类别
    Node    *next;

    Node();
};


/******************************************************************
    Examinee 类作为单链表类，用以实现相关操作
******************************************************************/
class Examinee
{
public:
    void    Creat   ( Examinee &L );
    void    Insert  ( Examinee &L );
    void    Delete  ( Examinee &L );
    void    Update  ( Examinee &L );
    void    Lookup  ( Examinee &L );
    int     Length  ( Examinee &L );
    void    Print   ( Examinee &L );

    Examinee();
```

```cpp
    private:
        Node    *head;
};
```

1.3.2.3 程序清单

```cpp
#include "Examinee.h"

/***************************************************************************
    Node 类的构造函数，初始化结点数据
***************************************************************************/
Node::Node()
{
    examnum = 0;
    name = '?';
    sex = '?';
    age = 0;
    type = '?';
    next = NULL;
}


/***************************************************************************
    Examinee 类的构造函数，初始化首结点
***************************************************************************/
Examinee::Examinee()
{
    head = NULL;
}


/***************************************************************************
    创建考生信息的线性表（单链表）
***************************************************************************/
void Examinee::Creat(Examinee &L)
{
    cout << "请输入考生人数：";
    int n;
    cin >> n;
    cout << "请依次输入考生的考号，姓名，性别，年龄及报考类别！" << '\n';
    Node *p, *s;
    for(int j = 0; j < n; j ++)
```

```cpp
        {
            p = new Node;
            long examnum;
            cin >> examnum;
            p->examnum = examnum;

            string name;
            cin >> name;
            p->name = name;

            string sex;
            cin >> sex;
            p->sex = sex;

            int age;
            cin >> age;
            p->age = age;

            string type;
            cin >> type;
            p->type = type;

            p->next = NULL;

            if (j == 0) {
                L.head = p;
                s = L.head;
            }
            else {
                s->next = p;
                s = p;
            }
        }
    }
}

/*******************************************************************
    输出线性表中所有的考生信息
*******************************************************************/
void Examinee::Print(Examinee &L)
{
```

```cpp
    Node *p = L.head;
    cout << '\n';
    cout << "考号" << '\t' << "姓名" << '\t' << "性别" << '\t';
    cout << "年龄" << '\t' << "报考类别" << '\n';
    while (p->next) {
        cout << p->examnum << '\t' << p->name << '\t' << p->sex << '\t';
        cout << p->age << '\t' << p->type << '\n';
        p = p->next;
    }
    cout << p->examnum << '\t' << p->name << '\t' << p->sex << '\t';
    cout << p->age << '\t' << p->type << '\n';
}


/************************************************************************
    查询线性表的长度
*************************************************************************/
int Examinee::Length(Examinee &L)
{
    Node *p = L.head;
    int i = 1;
    while (p->next) {
        i ++;
        p = p->next;
    }
    return i;
}


/************************************************************************
    在线性表的某个位置上插入考生信息
*************************************************************************/
void Examinee::Insert(Examinee &L)
{
    cout << "请输入你要插入的考生的位置：";
    int m;
    cin >> m;

    if (m > Examinee::Length(L) + 1) {
        cout << "你要插入的位置不存在，请重新输入！" << '\n';
        Examinee::Insert(L);
    }
```

```
else {
    Node *p, *s;

    p = new Node;
    s = L.head;
    cout << "请依次输入要插入的考生的考号，姓名，性别，年龄及报考类别！" << '\n';
    long examnum;
    cin >> examnum;
    p->examnum = examnum;

    string name;
    cin >> name;
    p->name = name;

    string sex;
    cin >> sex;
    p->sex = sex;

    int age;
    cin >> age;
    p->age = age;

    string type;
    cin >> type;
    p->type = type;

    if (m == 1) {
        p->next = L.head;
        L.head = p;
    }
    else {
        int i = 1;
        while (i != m - 1) {
            s = s->next;
            i ++;
        }
        p->next = s->next;
        s->next = p;
    }
}
```

}

/***
　　根据考号删除线性表中对应的学生信息
***/
void Examinee::Delete(Examinee &L)
{
　　cout << "请输入要删除的考生的考号：";
　　long examnum;
　　cin >> examnum;
　　Node *p = L.head;
　　Node *s;
　　int i = 1;
　　while (p->next) {
　　　　if (L.head->examnum == examnum) {
　　　　　　s = L.head;
　　　　　　L.head = s->next;
　　　　　　cout << "你删除的考生信息是：" << s->examnum << '\t' << s->name << '\t';
　　　　　　cout << s->sex << '\t' << s->age << '\t' << s->type << '\n';
　　　　　　break;
　　　　}
　　　　else if (p->next->examnum == examnum) {
　　　　　　s = p->next;
　　　　　　p->next = s->next;
　　　　　　cout << "你删除的考生信息是：" << s->examnum << '\t' << s->name << '\t';
　　　　　　cout << s->sex << '\t' << s->age << '\t' << s->type << '\n';
　　　　　　break;
　　　　}
　　　　else {
　　　　　　p = p->next;
　　　　　　i ++;
　　　　}
　　}
　　if (i > Examinee::Length(L)) {
　　　　cout << "你要删除的考生不存在，请重新输入考号！" << '\n';
　　　　Examinee::Delete(L);
　　}
}

```
/******************************************************************
    根据考号查询线性表中的考生信息
*******************************************************************/
void Examinee::Lookup(Examinee &L)
{
    cout << "请输入要查找的考生的考号：";
    long examnum;
    cin >> examnum;
    Node *p = L.head;
    int i = 1;
    if (i > Examinee::Length(L)) {
        cout << "你要查找的考生不存在，请重新输入考号！" << '\n';
        Examinee::Lookup(L);
    }
    while (p) {
        if (p->examnum == examnum) {
            cout << "考号" << '\t' << "姓名" << '\t' << "性别" << '\t';
            cout << "年龄" << '\t' << "报考类别" << '\n';
            cout << p->examnum << '\t' << p->name << '\t' << p->sex << '\t';
            cout << p->age << '\t' << p->type << '\n';
            break;
        }
        else {
            p = p->next;
            i ++;
        }
    }
    cout<< '\n';
}

/******************************************************************
    根据考号修改线性表中的考生信息
*******************************************************************/
void Examinee::Update(Examinee &L)
{
    cout << "请输入要修改信息的考生的考号：";
    long examnum;
    cin >> examnum;
    Node *p = L.head;
    int i = 1;
```

```cpp
        if (i > Examinee::Length(L)) {
            cout << "你要修改信息的考生不存在，请重新输入考号！" << '\n';
            Examinee::Lookup(L);
        }

        while (p) {
            if (p->examnum == examnum) {
            cout << "请依次输入修改后的考生的考号，姓名，性别，年龄及报考类别！" << '\n';
                long updatenum;
                cin >> updatenum;
                p->examnum = updatenum;

                string name;
                cin >> name;
                p->name = name;

                string sex;
                cin >> sex;
                p->sex = sex;

                int age;
                cin >> age;
                p->age = age;

                string type;
                cin >> type;
                p->type = type;
                break;
            }
            else {
                p = p->next;
                i ++;
            }
        }
    }
}

/*********************************************************************
        主函数，调用类 Examinee 的成员函数以实现相应功能
*********************************************************************/
void main()
```

```cpp
{
    cout << "首先请建立考生信息系统！" << '\n';
    Examinee L;
    L.Creat(L);
    L.Print(L);

    int j=-1;
    cout << "请选择您要进行的操作（1 为插入，2 为删除，3 为查找，4 为修改，5 为统计，0 为取消操作）";
    while(j)
    {
        cout << '\n'<<"请选择您要进行的操作：";
        cin >> j;
        switch(j){
        case 1:
            {
                L.Insert(L);
                L.Print(L);
                break;
            }
        case 2:
            {
                L.Delete(L);
                L.Print(L);
                break;
            }
        case 3:
            {
                L.Lookup(L);
                break;
            }
        case 4:
            {
                L.Update(L);
                L.Print(L);
                break;
            }
        case 5:
            {
                cout << "该考生系统人数为：" << L.Length(L) << '\n';
```

```
                break;
            }
        default:
            {
                cout << "该操作不存在，请输入正确操作！" << '\n';
                break;
            }
        }
    }
}
```

1.3.2.4 运行结果

```
首先请建立考生信息系统!
请输入考生人数: 3
请依次输入考生的考号，姓名，性别，年龄及报考类别!
1 stu1 女 20 软件设计师
2 stu2 男 21 软件开发师
3 stu3 男 20 软件设计师

考号      姓名      性别      年龄      报考类别
1         stu1      女        20        软件设计师
2         stu2      男        21        软件开发师
3         stu3      男        20        软件设计师
请选择您要进行的操作（1为插入，2为删除，3为查找，4为修改，5为统计，0为取消操作）

请选择您要进行的操作: 1
请输入你要插入的考生的位置: 4
请依次输入要插入的考生的考号，姓名，性别，年龄及报考类别!
4 stu4 女 21 软件测试师

考号      姓名      性别      年龄      报考类别
1         stu1      女        20        软件设计师
2         stu2      男        21        软件开发师
3         stu3      男        20        软件设计师
4         stu4      女        21        软件测试师
请选择您要进行的操作: 2
请输入要删除的考生的考号: 2
你删除的考生信息是: 2    stu2       男        21         软件开发师

考号      姓名      性别      年龄      报考类别
1         stu1      女        20        软件设计师
3         stu3      男        20        软件设计师
4         stu4      女        21        软件测试师
请选择您要进行的操作: 3
请输入要查找的考生的考号: 3
考号      姓名      性别      年龄      报考类别
3         stu3      男        20        软件设计师
```

1.3.3 约瑟夫生者死者游戏

1.3.3.1 游戏简介

约瑟夫生者死者游戏的大意是：30个旅客同乘一条船，因为严重超载，加上风高浪大，危险万分；因此船长告诉乘客，只有将全船一半的旅客投入海中，其余人才能幸免于难。无奈，大家只得同意这种办法，并议定30个人围成一圈，由第一个人开始，依次报数，数到第9人，便把他投入大海中，然后从他的下一个人数起，数到第9人，再将他投入大海，如此循环，直到剩下15个乘客为止。问哪些位置是将被扔下大海的位置。

1.3.3.2 设计思路

本游戏的数学建模如下：假设N个旅客排成一个环形，依次顺序编号1，2，…，N。从某个指定的第S号开始，沿环计数，每数到第M个人就让其出列，且从下一个人开始重新计数，继续进行下去。这个过程一直进行到剩下K个旅客为止。

本游戏要求用户输入的内容包括：
1. 旅客的个数，也就是N的值；
2. 离开旅客的间隔数，也就是M的值；
3. 所有旅客的序号作为一组数据要求存放在某种数据结构中。

本游戏要求输出的内容是包括：
1. 离开旅客的序号；
2. 剩余旅客的序号。

所以，根据上面的模型分析及输入输出参数分析，可以定义一种数据结构后进行算法实现。

○ **数据结构**

为了解决这一问题，可以用长度为N的数组作为线性存储结构，并把该数组看成是一个首尾相接的环形结构，那么每选中投入大海一个乘客，就要在该数组的相应位置进行一次删除操作，该单元以后的单元顺次前移一个单元。这样算法较为复杂，而且效率低，还要移动大量的元素。

用单循环链表解决这一问题，实现的方法相对要简单得多。首先要定义链表结点，单循环链表的结点结构与一般的结点结构完全相同，只是数据域用一个整数来表示位置；然后将它们组成具有N个结点的单循环链表。接下来从位置为S的结点开始数，数到第M个结点，就将此结点从循环链表中删去，然后再从删去结点的下一个结点开始数起，数到第 M 个结点，再将此结点删去，如此进行下去，直到剩下 K 个结点为止。

○**程序设计**

具体算法描述如下：

（1）创建含有 N 个结点的单循环链表；

（2）生者与死者的选择：

p 指向链表的第一个结点，初始 i 置为 1；

while（i<=K）

　{ 从 p 指向的结点沿链表前进 k-1 步；

　　删除第 k 个结点（q 所指向的结点）；

　　p 指向 q 的下一个结点；

　　输出其位置 q->data；

　　i 自增 1；

　}

（3）输出所有生者的位置。

○**头文件定义**

单循环链表类 Linklist 及其结点类 Node 在头文件 Joseph.h 中定义为：

```
/************************************************************************
    单循环链表的结点类
************************************************************************/
class Node
{
    friend class Linklist;
public:
    Node();
    int data;
    Node *next;
};
/************************************************************************
    单循环链表类
************************************************************************/
class Linklist
{
public:
    Linklist();
    void Creatlist(Linklist &L);
    int getLength(Linklist &L);

    Node * Head;
};
```

1.3.3.3 程序清单

```cpp
#include "Joseph.h"
/*********************************************************************
    Node 类的构造函数，初始化结点数据
*********************************************************************/
Node::Node()
{
    data = 0;
    next = NULL;
}

/*********************************************************************
    Linklist 类的构造函数，初始化首结点数据
*********************************************************************/
Linklist::Linklist()
{
    Head = NULL;
}

/*********************************************************************
    建立单循环链表
*********************************************************************/
void Linklist::Creatlist(Linklist &L)
{
    cout << "请输入生死游戏的总人数 N：";
    int n;
    cin >> n;
    Node *p;
    for (int i = n; i > 0; i --)
    {
        p = new Node;
        p->data = i;
        p->next = L.Head;
        L.Head = p;
    }
    while (p->next) {
        p = p->next;
    }
    p->next = L.Head;
```

}

/***
    获取单循环链表的长度
**/
```cpp
int Linklist::getLength(Linklist &L)
{
    Node *p=L.Head;
    int count=0;
    while (p->next!=L.Head) {
        count++;
        p = p->next;
    }
    count++;
    return count;
}
```

/***
    主函数，实现约瑟夫生死游戏
**/
```cpp
void main ()
{
    cout << "现有 N 人围成一圈，从第 S 个人开始依次报数，报 M 的人出局，";
    cout << "再由下一人开始报数，如此循环，直至剩下 K 个人为止" << '\n' << '\n';

    Linklist L;
    L.Creatlist(L);

    cout << "请输入游戏开始的位置 S：" << '\t' ;
    int s;
    cin >> s;

    cout << "请输入死亡数字 M：" << '\t';
    int m;
    cin >> m;

    cout << "请输入剩余的生者人数 K：" << '\t';
    int k;
    cin >> k;
```

29

```
cout<< '\n' ;

Node *p, *q, *r;
p = L.Head;
for(int i = 0; i < s - 1; i ++)              //找出 s 结点
{
    p = p->next;
}

int t = 1;
while ( k < L.getLength(L) ) {
    for(int j = 0; j < m - 1; j ++)          //报数，找出 m 结点
    {
        q = p;
        p = p->next;
    }
    if (p == L.Head) {                       //元素出列
        r = p;
        L.Head = p->next;
        q->next = p->next;
        p = p->next;
    }
    else
    {
        r = p;
        q->next = p->next;
        p = p->next;
    }
    cout << "第" << t << "个死者的位置是：" << '\t'<< r->data << '\n';
    t ++;
}

cout<< '\n' << "最后剩下：" << '\t' << L.getLength(L) << "人" <<endl;
cout<< "剩余的生者位置为：" << '\t';
p = L.Head;

while(p->next!=L.Head){
    cout<< p->data << '\t';
    p=p->next;
}
```

```
        if(p)
            cout<< p->data << '\t' ;
    cout<< '\n';
}
```

1.3.3.4 运行结果

```
现有N人围成一圈,从第S个人开始依次报数,报M的人出局,再由下一人开始报数,如此循
环,直至剩下K个人为止
请输入生死游戏的总人数N: 30
请输入游戏开始的位置S: 1
请输入死亡数字M: 9
请输入剩余的生者人数K: 15

第1个死者的位置是:       9
第2个死者的位置是:      18
第3个死者的位置是:      27
第4个死者的位置是:       6
第5个死者的位置是:      16
第6个死者的位置是:      26
第7个死者的位置是:       7
第8个死者的位置是:      19
第9个死者的位置是:      30
第10个死者的位置是:     12
第11个死者的位置是:     24
第12个死者的位置是:      8
第13个死者的位置是:     22
第14个死者的位置是:      5
第15个死者的位置是:     23

最后剩下:      15人
剩余的生者位置为:       1       2       3       4       10      11      13
14      15      17      20      21      25      28      29
Press any key to continue_
```

1.3.4 约瑟夫双向生死游戏

1.3.4.1 游戏简介

约瑟夫双向生死游戏是在约瑟夫生者死者游戏的基础上,正向计数后反向计数,然后再正向计数。具体描述如下:30 个旅客同乘一条船,因为严重超载,加上风高浪大,危险万分;因此船长告诉乘客,只有将全船一半的旅客投入海中,其余人才能幸免于难。无奈,大家只得同意这种办法,并议定 30 个人围成一圈,由第一个人开始,顺时针依次报数,数到第 9人,便把他投入大海中,然后从他的下一个人数起,逆时针数到第 5 人,将他投入大海,然后从他逆时针的下一个人数起,顺时针数到第 9 人,再将他投入大海,如此循环,直到剩下 15 个乘客为止。问哪些位置是将被扔下大海的位置。

1.3.4.2 设计思路

本游戏的数学建模如下:假设 n 个旅客排成一个环形,依次顺序编号 1, 2, …, n。从某个指定的第 1 号开始,沿环计数,数到第 m 个人就让其出列,然后从第 m+1 个人反向计

数到 m-w+1 个人，让其出列，然后从 m-w 个人开始重新正向沿环计数，再数 m 个人后让其出列，然后再反向数 w 个人后让其出列。这个过程一直进行到剩下 k 个旅客为止。

本游戏的要求用户输入的内容包括：

1. 旅客的个数，也就是 n 的值；
2. 正向离开旅客的间隔数，也就是 m 的值；
3. 反向离开旅客的间隔数，也就是 w 的值；
4. 所有旅客的序号作为一组数据要求存放在某种数据结构中。

本游戏要求输出的内容包括：

1. 离开旅客的序号；
2. 剩余旅客的序号。

根据上面的模型分析及输入输出参数分析，可以定义一种数据结构后进行算法实现。

○数据结构

约瑟夫双向生死游戏如果用单循环链表作为线性存储结构，就只能正向计数结点，反向计数比较困难，算法较为复杂，而且效率低。用双向循环链表解决这一问题，实现的方法相对要简单得多。

○程序设计

具体算法描述如下：

（1）创建含有 n 个结点的双向循环链表；

（2）生者与死者的选择：

p 指向链表的第一个结点，初始 i 置为 1；

while（i<=k）

  { 从 p 指向的结点沿链前进 m-1 步；

    删除第 m 个结点（q 所指向的结点）；

    p 指向 q 的下一个结点；

    输出其位置 q->data；

    i 自增 1；

    从 p 指向的结点沿链后退 w-1 步；

    删除第 w 个结点（q 所指向的结点）；

    p 指向 q 的上一个结点；

    输出其位置 q->data；

    i 自增 1；

  }

（3）输出所有生者的位置。

○头文件定义

双向循环链表类 Doublelist 及其结点类 Node 在头文件 DoubleJoseph.h 中定义为：

```
/***************************************************************
    双向循环链表的结点类
***************************************************************/
class Node
{
```

```cpp
    friend    class     Doublelist;
    friend    void      DoubleJoseph();
public:
    int       data;
    Node      *prior;
    Node      *next;
    Node();
};
```

/***
    双向循环链表类
***/

```cpp
class Doublelist
{
    friend    void      DoubleJoseph();
public:
    void      Creatlist    ( Doublelist &L );
    int       getLength    ( Doublelist &L );
    Doublelist();

private:
    Node *Head;
};
```

1.3.4.3 程序清单

```cpp
#include "DoubleJoseph.h"
/*************************************************************************
    Node 类的构造函数，初始化结点数据
*************************************************************************/
Node::Node()
{
    data = 0;
    prior = NULL;
    next = NULL;
}


/*************************************************************************
    Doublelist 类的构造函数，初始化首结点数据
*************************************************************************/
Doublelist::Doublelist()
```

```
{
    Head = NULL;
}
/****************************************************************
    建立双向循环链表
****************************************************************/
void Doublelist::Creatlist(Doublelist &L)
{
    cout << "请输入双向生死游戏的总人数 N：";
    int n;
    cin >> n;
    Node *p, *s;
    for(int i = 1; i <= n; i ++)
    {
        p = new Node;
        p->data = i;
        p->next = NULL;
        if (i == 1) {
            L.Head = p;
            p->prior = NULL;
            s = L.Head;
        }
        else
        {
            s->next = p;
            p->prior = s;
            s = s->next;
        }
    }
    p->next = L.Head;
    L.Head->prior = p;
}

/****************************************************************
    获取双向循环链表的长度
****************************************************************/
int Doublelist::getLength(Doublelist &L)
{
    Node *p=L.Head;
    int count=0;
```

```cpp
        while (p->next!=L.Head) {
            count++;
            p = p->next;
        }
        count++;
        return count;
}


/*********************************************************************
    实现约瑟夫双向生死游戏
*********************************************************************/
void DoubleJoseph()
{
    Doublelist L;
    L.Creatlist(L);

    cout << "请输入游戏开始的位置 S：";
    int s;
    cin >> s;

    cout << "请输入正向的死亡数字 M：";
    int m;
    cin >> m;

    cout << "请输入逆向的死亡数字 W：";
    int w;
    cin >> w;

    cout << "请输入剩余的生者人数 K：";
    int k;
    cin >> k;
    cout<< '\n';

    Node *p, *q, *r;
    p = L.Head;
    for(int i = 0; i < s - 1; i ++)                    //寻找 s 结点
    {
        p = p->next;
    }
    int t = 1;
```

```
while (k<L.getLength(L)) {
    if (t%2) {                                          //选择游戏方式
        for(int j = 0; j < m - 1; j ++)                 //报数，寻找 m 结点
        {
            q = p;
            p = p->next;
        }
        if (p == L.Head) {                              //元素出列
            r = p;
            L.Head = p->next;
            q->next = p->next;
            p->next->prior = q;
            p = p->next;
        }
        else
        {
            r = p;
            q->next = p->next;
            p->next->prior = q;
            p = p->next;
        }
        cout << "第" << t << "个死者的位置是：" << r->data << '\n';
        t ++;
    }
    else
    {
        for(int j = 0; j < w - 1; j ++)                 //报数，寻找 m 结点
        {
            q = p;
            p = p->prior;
        }
        if (p == L.Head) {                              //元素出列
            r = p;
            L.Head = p->prior;
            q->prior = p->prior;
            p->prior->next = q;
            p = p->prior;
        }
        else
        {
```

```cpp
                r = p;
                q->prior = p->prior;
                p->prior->next = q;
                p = p->prior;
            }
            cout << "第" << t << "个死者的位置是：" << r->data << '\n';
            t ++;
        }
    }

    cout<< '\n' << "最后剩下：" << '\t' << L.getLength(L) << "人" <<endl;
    cout<< "剩余的生者位置为：" << '\t';
    p = L.Head;

    while(p->next!=L.Head){
        cout<< p->data << '\t';
        p=p->next;
    }
    if(p)
        cout<< p->data << '\t' ;
    cout<< '\n';
}


/*************************************************************************
    主函数，调用 DoubleJoseph()函数
*************************************************************************/
void main()
{
    cout << "现有 N 人围成一圈，从第 S 个人开始依次报数，正向报 M 的人出局，再由下一人开始报数，";
    cout << "逆时针数到的第 W 人出局。再从他逆时针的下一个人数起，顺时针数 M 人。";
    cout << "如此循环，直到剩下 K 个乘客为止。"<< '\n' << '\n';

    DoubleJoseph();
}
```

1.3.4.4 运行结果

现有N人围成一圈，从第S个人开始依次报数，正向报M的人出局，再由下一人开始报数，逆时针数到的第W人出局。再从他逆时针的下一个人数起，顺时针数M人。如此循环，直到剩下K个乘客为止。
请输入双向生死游戏的总人数N: 30
请输入游戏开始的位置S: 1
请输入正向的死亡数字M: 9
请输入逆向的死亡数字W: 5
请输入剩余的生者人数K: 15

第1个死者的位置是：9
第2个死者的位置是：5
第3个死者的位置是：14
第4个死者的位置是：10
第5个死者的位置是：19
第6个死者的位置是：15
第7个死者的位置是：24
第8个死者的位置是：20
第9个死者的位置是：29
第10个死者的位置是：25
第11个死者的位置是：4
第12个死者的位置是：30
第13个死者的位置是：12
第14个死者的位置是：6
第15个死者的位置是：21

最后剩下：   15人
剩余的生者位置为：   1   2   3   7   8   11   13
16   17   18   22   23   26   27   28
Press any key to continue

1.4 巩固提高

设计一个通讯录管理系统，实现通讯录管理的几种操作功能，使用含有多个菜单项的控制台选择方式完成下列功能：通讯录链表的建立、通讯者结点的插入、通讯者结点的查询、通讯者结点的删除、通讯者结点的输出，以及退出管理系统。

○提示

本项目的实质是完成对通讯录的建立、查找、插入、修改、删除等功能。

项目在设计时应首先确定系统的数据结构，定义类的成员变量和成员函数；然后实现各成员函数以完成对数据操作的相应功能；最后完成主函数以验证各个成员函数的功能并得出运行结果。

第二章 栈和队列

栈和队列是两种重要的线性结构。从数据结构角度上看，栈和队列也是线性表，其特殊性在于栈和队列的基本操作是线性表操作的子集，它们是受限的线性表，因此，可称为限定性的数据结构。但从数据类型角度看，它们是和线性表大不相同的两类重要的抽象数据类型。由于它们广泛应用在各种软件系统中，因此在面向对象的程序设计中，它们是多型数据类型。本章通过勇闯迷宫游戏、N 皇后问题和停车场管理三个项目来学习栈和队列的定义、表示方法及其实现。

2.1 实践目的和要求

2.1.1 实践目的

1）掌握栈和队列的思想及其存储实现；
2）掌握栈和队列的常见算法的程序实现；
3）掌握上机实现栈和队列的基本方法；
4）熟悉并能实现栈的定义和初始化、栈的判空、入栈、出栈、取栈顶元素、置栈空等操作；
5）熟悉并能实现队列的定义和初始化、队列判空、出队列、入队列等操作；
6）掌握用队列解决实际应用问题。

2.1.2 实践要求

1）掌握本章实践的算法。
2）上机运行本章的程序，保存和打印出程序的运行结果，并结合程序进行分析。
3）进行栈的基本操作时要注意栈"后进先出"的特性。
4）进行队列的基本操作时要注意队列"先进先出"的特性。
5）重点理解栈、队列的算法思想，能够根据实际情况选择合适的存储结构。
6）栈和队列的算法是后续实践的基础（树、图、查找、排序等），要注意对这部分的常见算法的理解和掌握。

2.2 基本概念

2.2.1 栈

2.2.1.1 栈的定义

栈（stack）是一种操作受限即只允许在一端（表尾端）进行插入和删除操作的线性表。表中允许进行插入和删除的一端（表尾端）称为栈顶（top）。栈顶的当前位置是动态的，由一个称为栈顶指针的位置指示器指示。表的另一端（表头端）称为栈底（bottom）。

若有栈 $S=(s_1, s_2, \cdots, s_n)$，则 s_1 为栈底结点，s_n 为栈顶结点。向栈中插入新元素即栈的插入操作通常称为入栈或进栈（push），新元素入栈后就成为新的栈顶元素；而从栈中删除栈顶元素即栈的删除操作则称为出栈或退栈（pop），栈顶元素出栈后其直接前驱成为新的栈顶元素。当栈中无数据元素时，称为空栈。

根据栈的定义可知，栈顶元素总是最后入栈的，因而是最先出栈；栈底元素总是最先入栈的，因而也是最后出栈。这种表是按照后进先出（LIFO，last in first out）的原则组织数据的，因此，栈也被称为"后进先出"的线性表。"后进先出"即后进栈的元素先弹出是栈的主要特点。

图 2.1 是一个栈的示意图，通常用指针 top 指示栈顶的位置，用指针 bottom 指向栈底。栈顶指针 top 动态反映栈顶的当前位置。

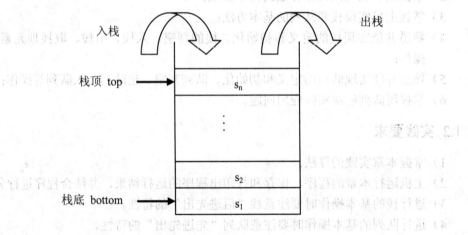

图2.1 栈的示意图

2.2.1.2 栈的存储结构

栈是一种特殊的线性表，因此与线性表类似，栈也可以采用顺序存储结构和链式存储结构。

○栈的顺序存储结构

栈的顺序存储结构即利用一组地址连续的存储单元依次存放自栈底到栈顶的数据元素，并用一个整型变量 top 指向当前的栈顶即栈顶元素的当前位置，这种形式的栈也称为顺序栈。因此，我们可以使用一维数组来作为栈的顺序存储空间，另外再设一个指针 top 指向栈顶元素的当前位置，以数组小下标的一端作为栈底，通常以 top=0 表示空栈，在元素进栈时指针 top 不断地加 1，当 top 等于数组的最大下标值时则栈满。

○栈的链式存储结构

栈也可以采用链式存储结构表示，这里采用单链表来实现，这种结构的栈简称为链栈。规定链栈的所有操作都在单链表的表头进行，即在一个链栈中，栈底就是链表的最后一个结点，而栈顶总是链表的第一个结点。因此，新入栈的元素即为链表新的第一个结点，只要系统还有存储空间，就不会有栈满的情况发生，即链栈不存在栈满上溢的情况，这是链栈的优点。一个链栈可由栈顶指针 top 唯一确定（不需再为链表设置头指针），top 为 NULL 表示是一个空栈。图 2.2 给出了链栈中数据元素与栈顶指针 top 的关系。

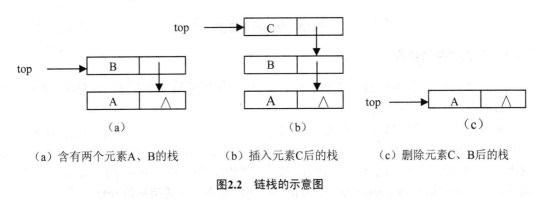

(a) 含有两个元素A、B的栈　　(b) 插入元素C后的栈　　(c) 删除元素C、B后的栈

图2.2　链栈的示意图

若程序中同时需两个以上的栈，则最好采用链式存储结构。

2.2.1.3 栈的基本运算

1) InitStack（&S）初始化：构造一个新的空栈。
2) StackEmpty（S）判空：若栈 S 为空栈，则返回 TRUE；否则，返回 FALSE。
3) Push（&S，e）入栈：在栈 S 的顶部插入元素 e，若栈满，则返回 FALSE；否则，返回 TRUE。
4) Pop（&S，&e）出栈：若栈 S 不空，则由 e 返回栈顶元素，并从栈中删除该元素；否则，返回空元素 NULL。
5) GetTop（s）取栈顶元素：若栈 S 不空，则返回栈顶元素；否则返回空元素 NULL。
6) ClearStack（&S）清空：置栈 S 为空栈。

2.2.2 队列

2.2.2.1 队列的定义

队列简称为队，也是一种操作受限的线性表，即只允许在表的一端进行插入而在表的另一端进行删除。允许进行插入的一端称为队尾（rear），允许进行删除的一端称为队头或队首（front）。若有队列 $Q=\{q_1, q_2, \cdots, q_n\}$，则队头元素为 q_1，队尾元素为 q_n。向队列中插入新元素即队列的插入操作通常称为入队或进队，新元素进入队列后就成为新的队尾元素；而从队列中删除队头元素即队列的删除操作则称为出队或离队，队头元素出队后其直接后继成为新的队头元素。当队列中无数据元素时，称为空队列。

根据队列的定义可知，队头元素总是最先进队列的，也总是最先出队列的；队尾元素总是最后进队列因而也是最后出队列的。这种表是按照先进先出（FIFO，first in first out）的原则组织数据的，因此，队列也被称为"先进先出"表。"先进先出"（即先进队的元素先出队）是队列的主要特点。

图 2.3 是一个队列的示意图，通常用指针 front 指示队头的位置，用指针 rear 指向队尾。

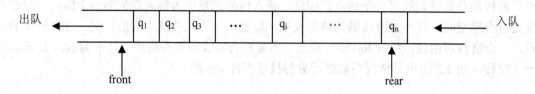

图2.3　队列的示意图

2.2.2.2 队列的存储结构

队列是一种特殊的线性表，它与线性表类似，也可以采用顺序存储结构和链式存储结构。

○队列的顺序存储结构

队列的顺序存储结构即利用一组地址连续的存储单元依次存放队列中的数据元素并用两个整型变量 rear 和 front 分别指向队头和队尾，这种形式的队列可以简称为顺序队列。一般情况下，我们使用一维数组来作为队列的顺序存储空间，另外再设立两个指示器：一个为指向队头元素位置的指示器 front，另一个为指向队尾的元素位置的指示器 rear。

在 C/C++语言中，数组的下标是从 0 开始的，因此为了算法设计的方便，在此我们约定：初始化队列时令 front=rear=-1，当插入新的数据元素时，尾指示器 rear 加 1，而当队头元素出队列时，队头指示器 front 加 1。另外还约定，在非空队列中，头指示器 front 总是指向队列中实际队头元素的前面一个位置，而尾指示器 rear 总是指向队尾元素。

图 2.4 给出了队列中头尾指针的变化状态。

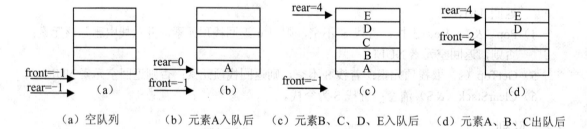

(a) 空队列　　　(b) 元素A入队后　　(c) 元素B、C、D、E入队后　　(d) 元素A、B、C出队后

图2.4　队列的顺序存储结构

○循环队列

在顺序队列中，当队尾指针已经指向了队列的最后一个位置时，此时若有元素入列，就会发生"溢出"。如图 2.4（c）所示的队列空间已满，若再有元素入列，则为溢出；在图 2.4（d）中，虽然队尾指针已经指向最后了一个位置，但事实上队列中还有三个空位置。也就是说，队列的存储空间并没有满，但队列却发生了溢出，我们称这种现象为假溢出。解决这个问题可采用如下方法：将顺序队列的存储空间假想为一个环状的空间，如图 2.5 所示，其中 MAXNUM 表示队列的最大长度。

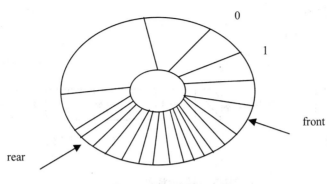

图2.5 循环队列示意图

这时，队列的0号位置（队列的第一个位置）紧接在队列的MAXNUM-1号位置（队列的最后一个位置）的后面，即首尾相接。当发生假溢出时，将新元素插入到0号位置上，这样作，虽然物理上队尾在队头之前，但逻辑上队头仍然在前。入队和出队仍按"先进先出"的原则进行，这就是循环队列。很显然，这种方法操作效率高，空间的利用率也很高。

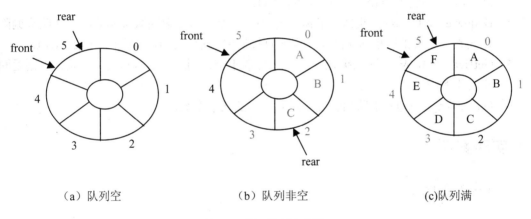

(a) 队列空　　　　　　　　(b) 队列非空　　　　　　　　(c) 队列满

图2.6 循环队列示意图

图2.6显示了循环队列在长度为6时的三种状态。图2.6（a）为队列空时，有front==rear；图2.6（c）为队列满时，也有front==rear；因此仅凭front==rear不能判定队列是空还是满，这是大家在使用循环队列时应该注意的一点。为了区分循环队列是空还是满，我们可以设定一个标志位s：s=0时为空队列，s=1时队列非空。

○队列的链式存储结构

队列也可以采用链式存储结构表示，这里采用单链表来实现，这种结构的队列简称为链队。规定只允许在单链表的表头进行删除操作和在单链表的表尾进行插入操作，即在一个链队中，队头就是链表的第一个结点，而队尾总是链表的最后一个结点。因此，新入队的元素即为链队新的最后一个结点，只要系统还有存储空间，就不会有队满的情况发生，即链队不存在队列满上溢的情况，这是链队的优点。在一个链队列中需设定两个指针（队头指针front和队尾指针rear）分别指向队列的队头结点和队尾结点。为了操作的方便，和线性链表一样，我们也给链队添加一个头结点，并设定队头指针指向头结点。因此，空队列的判定条件就是队头指针和队尾指针都指向头结点。

图 2.7 为链队示意图。

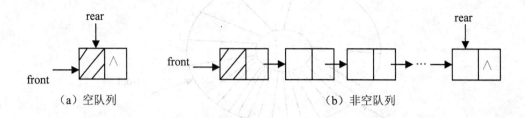

图2.7 链队示意图

2.2.2.3 队列的基本运算

1) InitQueue（&Q）初始化：初始化一个新的空队列 Q。
2) QueueEmpty（Q）队列判空：若队列 Q 为空，则返回 TRUE；否则，返回 FALSE。
3) EnQueue（&Q，e）入队：在队列 Q 的尾部插入元素 e，使元素 e 成为新的队尾。若队列满，则返回 FALSE；否则，返回 TRUE。
4) Dequeue（&Q，&e）出队：若队列 Q 不空，则通过 e 返回队头元素，并从队头删除该元素，队头指针指向原队头元素的直接后继元素；否则，返回空元素 NULL。
5) GetHead（Q，&e）取队头元素：若队列 Q 不空，则通过 e 返回队头元素；否则返回空元素 NULL。
6) QueueLength（Q）求队列长度：返回队列的元素个数。

2.3 实践案例

2.3.1 勇闯迷宫游戏

2.3.1.1 游戏简介

迷宫只有两个门，一个门叫入口，另一个门叫出口。一个骑士骑马从入口走进迷宫，迷宫中设置很多障碍，骑士需要在迷宫中寻找通路以到达出口。

2.3.1.2 设计思路

迷宫问题的求解过程可以采用回溯法即在一定的约束条件下试探地搜索前进，若前进中受阻，则及时回头纠正错误另择通路继续搜索的方法。从入口出发，按某一方向向前探索，若能走通（未走过的），即某处可达，则到达新点，否则试探下一方向；若所有的方向均没有通路，则沿原路返回前一点，换下一个方向再继续试探，直到所有可能的通路都探索到，或找到一条通路，或无路可走又返回到入口点。在求解过程中，为了保证在到达某一点后不能向前继续行走（无路）时，能正确返回前一点以便继续从下一个方向向前试探，则需要在试探过程中保存所能够到达的每一点的下标及从该点前进的方向，当找到出口时试探过程就结束了。

○数据结构

可以用二维数组表示二维迷宫中各个点是否有通路。在二维迷宫里面,从出发点开始,每个点按四邻域计算,按照右、上、左、下的顺序搜索一下落脚点,有路则走,无路即退回前点再从下一个方向搜索,即可实现回溯。

迷宫问题是栈应用的一个典型例子。通过前面分析,我们知道在试探过程中为了能够沿着原路逆序回退,就需要一种数据结构来保存试探过程中曾走过的点的下标及从该点前进的方向,在不能继续走下去时就要退回前一点继续试探下一个方向,栈底元素是入口,栈顶元素是回退的第一站,也即后走过的点先退回,先走过的点后退回,与栈的"后进先出,先进后出"特点一致,故在该问题求解的程序中可以采用栈这种数据结构。

○程序设计

栈可用链式结构实现。在迷宫有通路时,栈中保存的点逆序连起来就是一条迷宫的通路,否则栈中没有通路。

○头文件定义

位置类 Coordinate,链栈的结点类 Node 和链栈类 LinkStack 在头文件 Maze.h 中的具体定义为:

```
/*****************************************************************
    位置类
*****************************************************************/
class Coordinate
{
    friend    class    Node;
    friend    class    LinkStack;
    friend    bool    mazePath( char **maze, int m );
public:
    Coordinate();
    int   row;
    int   col;
};

/*****************************************************************
    链栈的结点类
*****************************************************************/
class Node
{
    friend    class    LinkStack;
public:
    Node();
    Coordinate    data;
    Node    *next;
};
```

```
/**************************************************************************
    链栈类
**************************************************************************/
class LinkStack
{
    friend bool mazePath(char **maze, int m);
public:
    void        Recursion       ( Node *P );
    void        PrintStack      ( LinkStack &L );
    bool        Empty           ( LinkStack &L );
    Coordinate  DeleteStack     ( LinkStack &L );
    void        AddStack        ( LinkStack &L, Coordinate N );
    LinkStack();

private:
    Node *top;
};
```

2.3.1.3 程序清单

```
#include "Maze.h"

/**************************************************************************
    位置类的初始化函数
**************************************************************************/
Coordinate::Coordinate()
{
    row = 1;
    col = 1;
}

/**************************************************************************
    结点类的初始化函数
**************************************************************************/
Node::Node()
{
    next = NULL;
}
```

```cpp
/************************************************************************
    链栈类的初始化函数
************************************************************************/
LinkStack::LinkStack()
{
    top = NULL;
}



/************************************************************************
    入栈
************************************************************************/
void LinkStack::AddStack(LinkStack &L, Coordinate N)
{
    Node *p = new Node;
    p->data = N;
    p->next = L.top;
    L.top = p;
}



/************************************************************************
    出栈
************************************************************************/
Coordinate LinkStack::DeleteStack(LinkStack &L)
{
    Node *p = L.top;
    Coordinate N = top->data;
    L.top = p->next;
    delete p;
    return N;
}



/************************************************************************
    判断栈是否为空
************************************************************************/
bool LinkStack::Empty(LinkStack &L)
{
    return (top == NULL) ? 1:0;
}
```

```
/********************************************************************
    递归输出栈的内容
********************************************************************/
void LinkStack::PrintStack(LinkStack &L)
{
    Node *p = new Node;
    p = L.top;
    LinkStack::Recursion(p);
    cout << '(' << p->data.row << ',' << p->data.col << ')' << " ---> ";
}


/********************************************************************
    递归函数
********************************************************************/
void LinkStack::Recursion(Node *P)
{
    if (P->next) {
        P = P->next;
        LinkStack::Recursion(P);
        cout << '(' << P->data.row << ',' << P->data.col << ')' << " ---> ";
    }
}



/********************************************************************
    实现迷宫算法,返回值判断是否到达迷宫出口
********************************************************************/
bool mazePath(char maze[7][7], int m)
{
    LinkStack path;

    //初始化偏移变量
    Coordinate move[4];
    move[0].row = 0; move[0].col = 1;      //向右移动
    move[1].row = 1; move[1].col = 0;      //向下移动
    move[2].row = 0; move[2].col = -1;     //向左移动
    move[3].row = -1; move[3].col = 0;     //向上移动

    //在迷宫周围增加一圈障碍物
    for(int i = 0; i <= m + 1; i ++)
```

```
    {
        maze[i][0] = maze[i][m + 1] = '#';        //左和右
        maze[0][i] = maze[m + 1][i] = '#';        //底和顶
    }

    Coordinate now;
    now.row = 1;
    now.col = 1;
    maze[1][1] = 'x';                              //设置"脚印"，阻止返回入口
    int choose = 0;
    int lastchoose = 3;

    //搜索一条路径
    while (now.row != m || now.col != m) {
        int newr, newc;

        while (choose <= lastchoose) {            //依次选择路径
            newr = now.row + move[choose].row;
            newc = now.col + move[choose].col;
            if (maze[newr][newc] == '0') {
                break;
            }
            choose ++;                             //进行下一个选择
        }

        if (choose <= lastchoose) {               //若找到相邻位置则移动到 maze[newr][newc]
            path.AddStack(path, now);
            now.row = newr;
            now.col = newc;

            maze[newr][newc] = 'x';               //设置"脚印"，以阻止再次访问
            choose = 0;
        }

        else {                                    //若没有找到相邻位置则回溯
            if (path.Empty(path)) {
                return false;
            }

            Coordinate Next;
```

```
            Next = path.DeleteStack(path);
            if (Next.row == Next.col) {
                choose = 2 + Next.col - now.col;
            }
            else {
                choose = 3 + Next.row - now.row;
            }
            now = Next;
        }
    }

    //显示迷宫地图
    cout << "迷宫地图： " << '\n' <<endl;
    for(int col = 0; col < 7; col++)
        cout<< '\t' << col << "列" ;
    cout<< '\n' ;
    for(int mi = 0; mi < 7; mi++)
    {
        cout<< mi << "行" ;
        for(int mj = 0; mj < 7; mj++)
            cout<< '\t' << maze[mi][mj] ;
        cout<< '\n' << '\n' ;
    }
    cout << '\n' ;

    cout << "迷宫路径： " <<    '\n' <<endl;
    path.PrintStack(path);
    cout << '(' << m << ',' << m << ')' << '\n' << '\n';

    return true;                        //到达迷宫出口
}
/********************************************************************
    主函数，调用 mazePath()函数
********************************************************************/
void main()
{
    int m = 5;
    char maze[7][7];
    for(int i = 1; i <= m; i ++)                //初始化迷宫数组
    {
```

50

```
            for(int j = 1; j <= m; j ++)
            {
                maze[i][j] = '0';
            }
        }
        maze[1][2] = '#';                    //设置迷宫中的障碍物
        maze[2][2] = '#';
        maze[2][4] = '#';
        maze[2][5] = '#';
        maze[3][4] = '#';
        maze[4][2] = '#';
        maze[5][2] = '#';
        maze[5][4] = '#';

        mazePath(maze, m);
    }
```

2.3.1.4 运行结果

2.3.2 N 皇后问题

2.3.2.1 问题简介

八皇后问题是一个古老而著名的问题，是回溯算法的典型例题。该问题是十九世纪著名的数学家高斯在 1850 年提出的：在 8×8 格的国际象棋棋盘上，安放八个皇后，要求没有一个皇后能够"吃掉"任何其他一个皇后，即任意两个皇后都不能处于同一行、同一列或同一条对角线上，求解有多少种摆法。

高斯认为有 76 种方案。1854 年在柏林的象棋杂志上不同的作者发表了 40 种不同的解，后来有人用图论的方法得出结论，有 92 种摆法。

本实验拓展到了 N 皇后问题，即皇后个数由用户输入。

2.3.2.2 设计思路

八皇后在棋盘上分布的各种可能的格局数目非常大，约等于 2^{32} 种，但是，可以将一些明显不满足问题要求的格局排除掉。由于任意两个皇后不能同行，即每一行只能放置一个皇后，因此将第 i 个皇后放置在第 i 行上。这样在放置第 i 个皇后时，只要考虑它与前 i-1 个皇后处于不同列和不同对角线位置上即可。

解决该问题采用回溯法。首先将第一个皇后放于第一行第一列，然后依次在下一行上放置下一个皇后，直到八个皇后全放置安全。在放置每一个皇后时，都依次对每一列进行检测，首先检测放在第一列是否与已放置的皇后冲突，如不冲突，则将皇后放置在该列，否则，选择该行的下一列进行检测。如整行的八列都冲突，则回到上一行，重新选择位置，依次类推。

○ **数据结构**

八皇后问题是栈应用的另一个典型例子。通过前面分析，我们知道在对八个皇后的位置进行试探的过程中，可能遇到在某一行上的所有位置都不安全的情况，这时就要退回到上一行，重新摆放在上一行放置的皇后。为了能够对上一行的皇后继续寻找下一个安全位置，就必须记得该皇后目前所在的位置，即需要一种数据结构来保存从第一行开始的每一行上的皇后所在的位置。在放置进行不下去时，已经放置的皇后中后放置的皇后位置先被纠正，先放置的皇后后被纠正，与栈的"后进先出，先进后出"特点一致，故在该问题求解的程序中可以采用栈这种数据结构。在八个皇后都放置安全时，栈中保存的数据就是八个皇后在八行上的列位置。

○ **程序设计**

用二维数组表示 N×N 格的国际象棋棋盘。栈可用顺序结构实现。本实验的栈结构设计较为巧妙，用 x[k]作为栈的顺序存储方式，直接存储第 i 个皇后放置的位置（行数）。

○ **头文件定义**

皇后类 Queen 在头文件 Queen.h 中的具体定义为：

```
/******************************************************************
    皇后类
******************************************************************/
class Queen
{
    friend    int nQueen ( int n );
public:
    void      Print ( int *x );
    void      Backtrack ( );
    bool      Place ( int  k );
```

```cpp
private:
    int     n;              //皇后个数
    int     *x;             //当前解
    int     sum;            //当前已找到的可行方案数
};
```

2.3.2.3 程序清单

```cpp
#include <math.h>
#include "Queen.h"
/***************************************************************
    判断同一行、列、斜线是否有其他皇后
****************************************************************/
bool Queen::Place(int k)
{
    for (int j = 1; j < k; j ++)
    {
        if ((abs(k - j) == abs(x[j] - x[k])) || (x[j] == x[k])) {
            return false;
        }
    }
    return true;
}


/***************************************************************
    进行迭代回溯
****************************************************************/
void Queen::Backtrack()
{
    x[1] = 0;
    int k = 1;
    while (k > 0) {
        x[k] += 1;
        while ((x[k] <= n) && !(Queen::Place(k))) {
            x[k] += 1;
        }
        if (x[k] <= n) {
            if (k == n) {
                sum ++;
                Queen::Print(x);
            }
```

```cpp
            else
            {
                k ++;
                x[k] = 0;
            }
        }
        else
            k --;
    }
}

/***********************************************************************
    输出结果
***********************************************************************/
void Queen::Print(int *x)
{
    for(int i = 1; i <= n; i ++)
    {
        for(int j = 1; j <= n; j ++)
        {
            if (i == x[j]) {
                cout << 'X' << ' ';
            }
            else
                cout << '0' << ' ';

        }
        cout << '\n';
    }
    cout << '\n';
}

/***********************************************************************
    问题求解
***********************************************************************/
int nQueen(int n)
{
    Queen X;
    X.n = n;
    X.sum = 0;
```

```cpp
    int *p = new int [n + 1];
    for(int i = 0; i <= n; i ++)
            p[i] = 0;
    X.x = p;
    X.Backtrack();
    delete [] p;
    return X.sum;
}
```

```
/*************************************************************************
    主函数，调用 nQueen()函数
*************************************************************************/
void main()
{
    cout << "现有 N×N 的棋盘，放入 N 个皇后，要求所有皇后不在同一行、列和同一斜线
上！" << '\n';

    cout << '\n' << "请输入皇后的个数：";
    int n;
    cin >> n;

    cout << '\n' << "皇后摆法：" << '\n' << '\n';

    int i = nQueen(n);
    cout << "共有" << i << "种解法！" << '\n';
}
```

2.3.2.4 运行结果

```
现有N×N的棋盘，放入N个皇后，要求所有皇后不在同一行、列和同一斜线上！
请输入皇后的个数：6

皇后摆法：

0 0 0 X 0 0
X 0 0 0 0 0
0 0 0 0 X 0
0 X 0 0 0 0
0 0 0 0 0 X
0 0 X 0 0 0

0 0 0 X 0 0
0 0 X 0 0 0
X 0 0 0 0 0
0 0 0 0 0 X
0 0 0 0 X 0
0 X 0 0 0 0

0 X 0 0 0 0
0 0 0 X 0 0
0 0 0 0 0 X
X 0 0 0 0 0
0 0 X 0 0 0
0 0 0 0 X 0

0 X 0 0 0 0
0 0 0 0 X 0
X 0 0 0 0 0
0 0 0 X 0 0
X 0 0 0 0 0
0 0 X 0 0 0

共有4种解法！
Press any key to continue_
```

2.3.3 停车场管理系统

2.3.3.1 系统简介

设停车场是一个可以停放 n 辆汽车的南北方向的狭长通道，且只有一个大门可供汽车进出。汽车在停车场内按车辆到达时间的先后顺序，依次由北向南排列（大门在最南端，最先到达的第一辆车停放在车场的最北端）。若车场内已停满 n 辆车，那么后来的车只能在门外的便道上等候。一旦停车场内某辆车要离开，在它之后进入的车辆必须先退出车场为它让路，待该辆车开出大门外，则排在便道上的第一辆车即可开入，其他车辆再按原次序进入车场。试为停车场编制按上述要求进行管理的模拟程序，要求程序输出车辆到达及离开后停车场与便道的停车情况。

2.3.3.2 设计思路

停车场的管理流程如下：

①当车辆要进入停车场时，检查停车场是否已满，如果未满则车辆进入停车场；如果停

车场已满，则车辆进入便道等候。

②当车辆要求离开时，先让在它之后进入停车场的车辆退出停车场为它让路，再让该车退出停车场，再检查在便道上是否有车等候，有车则让最先等待的那辆车进入停车场。之后，让路的所有车辆再按其原来进入停车场的次序进入停车场。

〇数据结构

由于这个停车场只有一个大门，当停车场内某辆车要离开时，在它之后进入的车辆必须先退出车场为它让路，先进停车场的后退出，后进车场的先退出，符合栈的"后进先出，先进后出"的操作特点，因此，可以用一个栈来模拟停车场。而当停车场满后，继续来到的其他车辆只能停在便道上，根据便道停车的特点，先排队的车辆先离开便道进入停车场，符合队列的"先进先出，后进后出"的操作特点，因此，可以用一个队列来模拟便道。排在停车场中间的车辆可以提出离开停车场，并且停车场内在要离开的车辆之后到达的车辆都必须先离开停车场为它让路，然后这些车辆依原来到达停车场的次序进入停车场，因此在前面已设的一个栈和一个队列的基础上，还需要有一个地方保存为了让路离开停车场的车辆，由于先退出停车场的后进入停车场，所以很显然保存让路车辆的场地也应该用一个栈来模拟。因此，本题求解过程中需用到两个栈和一个队列。栈以顺序结构实现，队列以链表结构实现。

〇程序设计

以栈模拟停车场，以队列模拟车场外的便道，按照从终端读入的输入数据序列进行模拟管理。每一组输入数据包括三个数据项：汽车到达或离去信息、汽车牌照号码以及到达或离去的时刻。对每一组输入数据进行操作后的输出信息为：若是车辆到达，则输出汽车在停车场内或便道上的停车位置；若是车辆离去，还可考虑增加输出汽车在停车场内停留的时间和应交纳的费用。

此实验中存储结构使用链表，更容易实现元素的插入与删除。

在停车场中使用栈的方式，在便道上使用队列。

〇头文件定义

```
/***********************************************************************
    车场停车信息
***********************************************************************/
class Parking_Space
{
    friend class Park;
    friend class Assistant;
    friend class Pavement;
    friend void storage(Park &L, Pavement &M);
    friend void take_out(Park &L, Assistant &N, Pavement &M);

public:
    int station;                        //停车状态
    string car_num;                     //车牌号
    time_t   parktime;                  //停车时间
    Parking_Space *next;                //下一车位
```

```cpp
        Parking_Space()
        {
            station = 0;
            car_num = '?';
            next = NULL;
        }
};

/************************************************************************
    模拟停车场的堆栈
*************************************************************************/
class Park
{
    friend void operate(Park &L, Assistant &N, Pavement &M);
    friend void storage(Park &L, Pavement &M);
    friend void take_out(Park &L, Assistant &N, Pavement &M);

public:
    static void Print(Park &L);
    static Parking_Space * Pop(Park &L);
    static void Put_in(Park &L, Parking_Space *p);
    static bool Full(Park &L);
    static int Length(Park &L);
    Park()
    {
        gate = NULL;
    }

private:
    Parking_Space *gate;
};

/************************************************************************
    辅助便道
*************************************************************************/
class Assistant
{
    friend void take_out(Park &L, Assistant &N, Pavement &M);
    friend void operate(Park &L, Assistant &N, Pavement &M);
```

```cpp
public:
    static bool Empty(Assistant &L);
    static Parking_Space * Pop(Assistant &L);
    static void Put_in(Assistant &L, Parking_Space *p);
    Assistant()
    {
        top = NULL;
    }

private:
    Parking_Space *top;
};

/************************************************************************
    模拟便道的队列
************************************************************************/
class Pavement
{
    friend void storage(Park &L, Pavement &M);
    friend void take_out(Park &L, Assistant &N, Pavement &M);
    friend void operate(Park &L, Assistant &N, Pavement &M);

public:
    static void Print(Pavement &L);
    static bool Empty(Pavement &L);
    static Parking_Space * Pop(Pavement &L);
    static void Put_in(Pavement &L, Parking_Space *p);
    Pavement()
    {
        head = NULL;
    }

private:
    Parking_Space *head;
};
```

2.3.3.3 程序清单

```
/************************************************************************
    打印停车场的车号
************************************************************************/
void Park::Print(Park &L)
{
    cout << "车位" << '\t' << "车牌号" << '\n';
    Parking_Space *p = L.gate;
    while (p) {
        cout << p->station << '\t' << p->car_num << '\n';
        p = p->next;
    }
}

/************************************************************************
    停车场的车队长度
************************************************************************/
int Park::Length(Park &L)
{
    Parking_Space *p = L.gate;
    int i = 0;
    while (p) {
        i ++;
        p = p->next;
    }
    return i;
}

/************************************************************************
    停车场是否已满
************************************************************************/
bool Park::Full(Park &L)
{
    if (Park::Length(L) == 10) {
        return true;
    }
    else
        return false;
}
```

/***
    进入停车场
***/
void Park::Put_in(Park &L, Parking_Space *p)
{
    p->station = Park::Length(L) + 1;
    p->next = L.gate;
    L.gate = p;
}

/***
    离开停车场
***/
Parking_Space * Park::Pop(Park &L)
{
    Parking_Space *p = L.gate;
    L.gate = p->next;
    return p;
}

/***
    辅助便道是否为空
***/
bool Assistant::Empty(Assistant &L)
{
    if (!L.top) {
        return true;
    }
    else
        return false;
}

/***
    车进入辅助便道
***/
void Assistant::Put_in(Assistant &L, Parking_Space *p)
{
    p->next = L.top;
    L.top = p;

61

}

/**
    车离开辅助便道
**/
Parking_Space * Assistant::Pop(Assistant &L)
{
    Parking_Space *p = L.top;
    L.top = p->next;
    return p;
}

/**
    便道是否为空
**/
bool Pavement::Empty(Pavement &L)
{
    if (!L.head) {
        return true;
    }
    else
        return false;
}

/**
    打印便道车号
**/
void Pavement::Print(Pavement &L)
{
    cout << "停在便道上的车的牌号依次是：";
    Parking_Space *p = L.head;
    while (p) {
        cout << p->car_num << '\t';
        p = p->next;
    }
    cout << '\n';
}

/***
    车进入便道
***/
void Pavement::Put_in(Pavement &L, Parking_Space *p)
{
    if (Pavement::Empty(L)) {
        p->next = NULL;
        L.head = p;
    }
    else {
        Parking_Space *s = L.head;
        while (s->next) {
            s = s->next;
        }
        p->next = NULL;
        s->next = p;
    }
}

/***
    车离开便道
***/
Parking_Space * Pavement::Pop(Pavement &L)
{
    Parking_Space *p = L.head;
    L.head = p->next;
    return p;
}

/***
    存车操作
***/
void storage(Park &L, Pavement &M)
{
    Parking_Space *p = new Parking_Space;
    cout << "请输入要存储的车牌号：";
    string car_num;
    cin >> car_num;
    p->car_num = car_num;
    if (!Park::Full(L)) {

```cpp
            Park::Put_in(L, p);
            cout << "车牌为" << p->car_num << "的车已进入停车场,所在的停车位为" << p->station << '\n';
            cout << "此时停车场内的汽车有：" << '\n';
            Park::Print(L);
        }
        else {
            cout << "停车场已满，此车将进入便道进行等候！" << '\n';
            Pavement::Put_in(M, p);
            cout << "此时停车场内的汽车有：" << '\n';
            Park::Print(L);
            Pavement::Print(M);
        }
    }
}

/************************************************************************
    取车操作
************************************************************************/
void take_out(Park &L, Assistant &N, Pavement &M)
{
    cout << "请输入要取出的汽车的牌号：";
    string car_num;
    cin >> car_num;
    Parking_Space *p = L.gate;
    while (p->car_num != car_num) {
        p = p->next;
        Parking_Space *s = Park::Pop(L);
        Assistant::Put_in(N, s);
    }
    Park::Pop(L);
    if (Pavement::Empty(M)) {
        cout << "便道没有车辆，空车位将由后面的汽车填补！" << '\n';
        while (!Assistant::Empty(N)) {
            Parking_Space *s = Assistant::Pop(N);
            Park::Put_in(L, s);
        }
        cout << "此时停车场内的汽车有：" << '\n';
        Park::Print(L);
    }
    else {
```

```cpp
        cout << "停在便道的第一辆车将进入空车位！" << '\n';
        Parking_Space *s = Pavement::Pop(M);
        Park::Put_in(L, s);
        while (!Assistant::Empty(N)) {
            Parking_Space *s = Assistant::Pop(N);
            Park::Put_in(L, s);
        }
        cout << "此时停车场内的汽车有：" << '\n';
        Park::Print(L);
        if (!Pavement::Empty(M)) {
            Pavement::Print(M);
        }
        else
            cout << "便道上的汽车已全部进入停车场！" << '\n';
    }
}

/************************************************************************
    操作选择函数，根据输入执行相应的操作
*************************************************************************/
void operate(Park &L, Assistant &N, Pavement &M)
{
    cout<< '\n' <<"请选择操作(A,D,E)   : ";
    char ch;
    cin >> ch;
    switch(ch) {
    case 'E':
        break;
    case 'A':
        {
            storage(L, M);
            operate(L, N, M);
            break;
        }
    case 'D':
        {
            take_out(L, N, M);
            operate(L, N, M);
            break;
        }
```

```
            default:
                {
                    cout << "输入错误，请选择正确的操作！" << '\n';
                    operate(L, N, M);
                    break;
                }
        }
    }

/**************************************************************************
    主函数，递归调用 operate()操作选择函数
**************************************************************************/
void main()
{
    Park L;
    Pavement M;
    Assistant N;

    cout<<"\n**                     停车场管理程序                      **"<<endl;
    cout<<"=========================================================="<<endl;
    cout<<"**                A --- 汽车 进 车场                       **"<<endl;
    cout<<"**                D --- 汽车 出 车场                       **"<<endl;
    cout<<"**                                                         **"<<endl;
    cout<<"**                E --- 退出    程序                       **"<<endl;
    cout<<"=========================================================="<<endl;
    cout<<"   请选择 :(A,D,E): ";

    operate(L, N, M);
}
```

2.3.3.4 运行结果

```
**              停车场管理程序              **
==========================================
**         A --- 汽车 进 车场              **
**         D --- 汽车 出 车场              **
**         E --- 退出 程序                 **
请选择:<A,D,E>:
请选择操作<A,D,E> : A
请输入要存储的车牌号: BJ9001
车牌为BJ9001的车已进入停车场,所在的停车位为1
此时停车场内的汽车有:
车位      车牌号
1         BJ9001

请选择操作<A,D,E> : A
请输入要存储的车牌号: BJ9002
车牌为BJ9002的车已进入停车场,所在的停车位为2
此时停车场内的汽车有:
车位      车牌号
2         BJ9002
1         BJ9001

请选择操作<A,D,E> : A
请输入要存储的车牌号: BJ9003
停车场已满，此车将进入便道进行等候!
此时停车场内的汽车有:
车位      车牌号
2         BJ9002
1         BJ9001
停在便道上的车的牌号依次是：BJ9003

请选择操作<A,D,E> : D
请输入要取出的汽车的牌号: BJ9001
停在便道的第一辆车将进入空车位!
此时停车场内的汽车有:
车位      车牌号
2         BJ9002
3         BJ9003
便道上的汽车已全部进入停车场!

请选择操作<A,D,E> : E
Press any key to continue
```

2.4 巩固提高

表达式计算是实现程序设计语言的基本问题之一，也是栈和队列应用的一个典型例子。请编程实现表达式求值，要求以字符序列的形式从终端输入语法正确的、不含变量的整数表达式，利用给定的算符优先关系，实现对算术四则混合运算表达式的求值，并演示在求值过程中运算符栈、操作数栈、输入字符和主要操作的变化过程。

○提示

人们在书写表达式时通常采用的"中缀"表达式，但是对于计算机处理来说，"后缀"表达式更为合适。因此，要用计算机来处理、计算算术表达式的问题，首先要解决的问题是如何将人们习惯书写的中缀表达式转换成计算机容易处理的后缀表达式。

第三章 串

计算机上的非数值处理的对象基本上是字符串数据。在较早的程序设计语言中，字符串是作为输入和输出的常量出现的。随着语言加工程序的发展，产生了字符串处理。这样，字符串也就作为一种变量类型出现在越来越多的程序设计语言中，同时也产生了一系列字符串的操作。字符串一般简称为串。在汇编和语言的编译程序中，源程序和目标程序都是字符串数据。在事务处理程序中，顾客的姓名和地址以及货物的名称、产地和规格等一般也是作为字符串处理的。又如信息检索系统、文字编辑程序、问答系统、自然语言翻译系统以及音乐分析程序等等，都是以字符串数据作为处理对象的。

然而，现今我们使用的计算机的硬件结构主要是反映数值计算的需要和效率，因此，在处理字符串数据时比处理整数和浮点数要复杂得多。而且，在不同类型的应用中，所处理的字符串具有不同的特点，要有效地实现字符串的处理，就必须根据具体情况使用合适的存储结构。本章通过关键字检索系统和四元线性方程组求解两个项目来学习串的定义、表示方法及其实现。

3.1 实践目的和要求

3.1.1 实践目的

1）理解字符串的基本概念、字符串与线性表的关系；
2）理解字符串的顺序存储和链式存储结构；
3）掌握上机实现串的基本方法；
4）理解和掌握串的基本操作：赋值、连接、求串长、求子串、求子串在主串中出现的位置、判断两个串是否相等、删除子串等。

3.1.2 实践要求

1）掌握本章实践的算法。
2）上机运行本章的程序，保存和打印出程序的运行结果，并结合程序进行分析。
3）按照你对串的操作需要，重新改写程序并运行，打印出文件清单和运行结果。

3.2 基本概念

3.2.1 串的定义

串是字符串（String）的简称。串是由零个或多个字符组成的有限序列。一般记作
$$S="c_1c_2\cdots c_n" \quad (n\geq 0)。$$

其中：S 为串名，用双引号括起来的字符序列是串的值简称为串值；双引号为串值的定界符，是串的标志，不是串的一部分，它的作用是为了避免将串与变量名或数值常量混淆。c_i（1≤i≤n）可以是大写或小写的英文字母、数字（0，1，2，…，9）、常用标点符号及空格符等。

字符串中字符的数目 n 称为串的长度。

长度为零的串称为空串（Null string），通常以两个相邻的双引号来表示空串，如：S=""，串定界符内的字符个数为零，空串也可以用φ表示。

仅由空格组成的串称为空格串，如：S="   "。若串中含有空格，在计算串长时，空格也应计入串的长度中，如：S="I am a student. "的长度为 15。

两个串相等当且仅当两个串的值相等，即两个串的长度相等并且各个对应位置上的字符也都相同。

一个串中任意 m（0≤m≤n）个连续的字符组成的子序列称为该串的子串，包含该子串的串称为主串。空串是任何串的子串。除串自身外的其余之中称为真子串。

一个字符在串序列中的序号称为该字符在串中的位置，子串在主串中的位置是以子串的第一个字符在主串中的位置来表示的。当一个字符在串中多次出现时，以该字符第一次在主串中出现的位置为该字符在串中的位置。

3.2.2 串的存储结构

对串的存储方式取决于我们对串所进行的运算，如果在程序设计语言中，串的运算只是作为输入或输出的常量出现，则此时只需存储该串的字符序列，这就是串值的存储。此外，一个字符序列还可赋给一个串变量，操作运算时通过串变量名访问串值。实现串名到串值的访问，在 C 语言中可以有两种方式：一是可以将串定义为字符型数组，数组名就是串名，串的存储空间分配在编译时完成，程序运行时不能更改。这种方式为串的静态存储结构。另一种是定义字符指针变量，存储串值的首地址，通过字符指针变量名访问串值，串的存储空间分配是在程序运行时动态分配的，这种方式称为串的动态存储结构。

串是数据对象为字符集的一种特殊的线性表，因此与线性表类似，串的存储结构也有多种表示方法：静态存储采用顺序存储结构，动态存储采用的是链式存储或堆存储结构。

3.2.2.1 串的静态存储结构

类似于线性表的顺序存储结构，用一组地址连续的存储单元存储串值的字符序列。由于一个字符只占 1 个字节，而现在大多数计算机的存储器地址是采用的字编址，一个字（即一个存储单元）占多个字节，因此顺序存储结构方式有两种：

○紧缩格式：即一个字节存储一个字符。这种存储方式可以在一个存储单元中存放多个字符，充分地利用了存储空间。但在串的操作运算时，若要分离某一部分字符时，则变得非常麻烦。

○非紧缩格式：这种方式是以一个存储单元为单位，每个存储单元仅存放一个字符。这种存储方式的空间利用率较低，如一个存储单元有 4 个字节，则空间利用率仅为 25%。但这种存储方式中不需要分离字符，因而程序处理字符的速度高。

串的顺序存储结构有两大不足之处：一是需事先预定义串的最大长度，这在程序运行前是很难估计的。二是由于定义了串的最大长度，使得串的某些操作受限，如串的连接运算等。

3.2.2.2 串的动态存储结构

我们知道，串的各种运算与串的存储结构有着很大的关系，在随机取子串时，顺序存储方式操作起来比较方便，而在对串进行插入、删除等操作时，就会变得很复杂。因此，有必要采用串的动态存储方式。

串的动态存储方式采用链式存储结构和堆存储结构两种形式。

（1）链式存储结构

类似于线性表的链式存储结构，采用链表方式存储串值字符序列。串的链式存储结构中每个结点包含字符域和结点链接指针域，字符域用于存放字符，指针域用于存放指向下一个结点的指针，因此，串可用单链表表示，此时串称为链串。

用单链表存放串，每个结点仅存储一个字符，因此，每个结点的指针域所占空间比字符域所占空间要大得多。为了提高空间的利用率，我们可以使每个结点存放多个字符，称为块链结构。通常将链串中每个结点所存储的字符个数称为结点大小（结点大小可以大于1）。

结点大小越大，则存储密度（存储密度=串值所占的存储位/实际分配的存储位）越大。但存储密度越大，一些操作如插入、删除、替换等越不方便，且可能引起大量字符移动，因此串的链式存储结构适合于在串基本保持静态使用方式时采用。结点大小越小如为1时，运算处理越方便，但存储密度下降。为简便起见，本章规定链串结点大小均为1。

（2）堆存储结构

堆存储结构的特点是，仍以一组空间足够大的、地址连续的存储单元存放串值字符序列，但它们的存储空间是在程序执行过程中动态分配的。每当产生一个新串时，系统就从剩余空间的起始处为串值分配一个长度和串值长度相等的存储空间。

在 C 语言中，存在一个称为"堆"的自由空间，由动态分配函数 malloc()分配一块实际串长所需的存储空间，如果分配成功，则返回这段空间的起始地址，作为串的基址。由 free()释放串不再需要的空间。

3.2.3 串的基本运算

串的基本运算有赋值、连接、求串长、求子串、求子串在主串中出现的位置、判断两个串是否相等、删除子串等。在本节中，我们尽可能以 C 语言的库函数表示其中的一些运算，若没有库函数，则需用自定义函数说明。

1) strcpy（&str1，str2）字符串拷贝（赋值）：把 str2 指向的字符串拷贝到 str1 中，返回 str1。C 语言的库函数和形参说明如下：char * strcpy（char * str1，char * str2）

2) strcat（str1，str2）字符串的连接：把字符串 str2 接到 str1 后面，str1 最后的结尾符′\0′被取消。返回 str1。库函数和形参说明如下：char * strcat（char * str1，char * str2）

3) strlen（str）求字符串的长度：统计字符串 str 中字符的个数（不包括′\0′），返回字符的个数，若 str 为空串，则返回值为 0。库函数和形参说明如下：unsigned int strlen（char *str）

4) strstr（str1，str2）子串的查询：找出子串 str2 在主串 str1 第一次出现的位置（不包括子串 str2 的结尾符），返回该位置的指针，若找不到，返回空指针 NULL。库函数和形参说明如下：char * strstr（char * str1，char * str2）

5) strcmp（str1，str2）字符串的比较：比较两个字符串 str1、str2。若 str1＜str2，则返

回负数；若 str1＞str2，则返回正数；若 str1＝str2，则返回 0。库函数和形参说明如下：int strcmp（char * str1，char * str2）

6）substr（str1，str2，m，n）求子串：在字符串 str1 中，从第 m 个字符开始，取 n 个长度的子串 str2；若 m＞strlen（str）或 n≤0，则返回空值 NULL。自定义函数和形参说明如下：int strstr（char * str1，char *str2，int m，int n）

7）delstr（str，m，n）字符串的删除：在字符串 str 中，删除从第 m 个字符开始的 n 个长度的子串。自定义函数和形参说明如下：void delstr（char *str，int m，int n）

8）insstr（str1，m，str2）字符串的插入：在字符串 str1 第 m 个位置之前开始，插入字符串 str2。返回 str1。自定义函数和形参说明如下：void insstr（char *str1，int m，char *str2）

对字符串的置换可以通过求串长、删除子串及字符串的连接等基本运算来实现。

3.3 实践案例

3.3.1 关键字检索系统

3.3.1.1 系统简介

建立一个文本文件，文件名由用户用键盘输入，输入一个不含空格的关键字，统计输出关键字在文本中的出现次数。

3.3.1.2 设计思路

本项目的设计要求可以分为两个部分实现：首先建立一个文本文件，文件名由用户用键盘输入；然后输入一个不含空格的关键字，统计输出该单词在文本中的出现次数。

○ **数据结构**

如果在程序设计语言中，串只是作为输入或输出的常量出现，则只需存储此串的串值，即字符序列即可。但在多数非数值处理的程序中，串也以变量的形式出现。串有三种机内表示方法：定长顺序存储表示、堆分配存储表示和串的块链存储表示。

定长顺序存储表示类似于线性表的顺序存储结构，用一组地址连续的存储单元存储串值的字符序列。在串的定长顺序存储结构中，按照预定义的大小，为每个定义的串变量分配一个固定长度的存储区，则可用定长数组描述。

本项目主要实现子串的检索操作，所以用定长顺序存储表示比较简单易行。

○ **程序设计**

1. 文件建立的实现过程

（1）建立一个文本文件，文件名由用户用键盘输入；

（2）定义一个字符变量；

（3）循环读入文本行，写入文本文件，其过程如下：

while（不是文件输入结束）｛

读入一文本行至串变量；

字符变量写入文件；

输入是否结束输入标志；
        }
    （4）关闭文件。
    2. 关键字计数的实现思路
    该功能需要用到模式匹配算法，逐行扫描文本文件。匹配一个，计数器加1，直到整个文件扫描结束；然后输出关键字出现的次数。
    串是非数值处理中的主要对象，如在信息检索、文本编辑、符号处理等许多领域，得到越来越广泛的应用。在串的基本操作中，在主串中查找模式串的模式匹配算法是文本处理中最常用、最重要的操作之一，称为模式匹配或串匹配，就是求子串在主串中首次出现的位置。朴素模式匹配算法的基本思路是将给定字串与主串从第一个字符开始比较，找到首次与子串完全匹配的子串为止，并记住该位置。但为了实现统计子串出现的个数，不仅需要从主串的第一个字符位置开始比较，而且需要从主串的任一位置检索匹配字符串。
    其实现过程如下：
    （1）输入要检索统计的关键字；
    （2）循环读文本文件，读入一行，将其送入定义好的串中，并求该串的实际长度，调用串匹配函数进行计数。具体描述如下：
        while（不是文件结束）{
            读入一行并到串中；
            模式匹配函数计数；
        }
    （3）关闭文件，输出统计结果。

3.3.1.3 程序清单

```
#include "iostream"
#include "fstream"
#include "string"
using namespace std;

extern char * filename = new char[];

/******************************************************************
    新建一个文件用来保存输入的英文
******************************************************************/
void input()
{
    cout<<"关键字检索系统"<<endl;

    cout << "请输入文件名："；
    cin >> filename;
    cout<<"\n 请输入一段英文："<<endl;
```

```cpp
    ofstream outfile(filename,ios::out);

    char ch;
    while (ch!='^' && ! cin.eof() && cin.get(ch)) {
        outfile << ch;
    }
    cout<<"本段文本已保存在文本文件\"" << filename << "\"中。 " <<endl;
    outfile.close();
}

/************************************************************************
    在源文件中检索相应的关键字
************************************************************************/
void search()
{
    int i = 0, j = 0;
    int number = 0;
    char ch;
    cout << "\n 请输入要检索的关键字：\t";
    char *key = new char[];
    cin >> key;
    int lenk = 0;
    while (key[lenk]) {
        lenk ++;
    }

    char *mod = new char[];
    int lenm = 0;

    cout<<"显示源文件\" " << filename << " \" ： " << endl;

    ifstream infile(filename, ios::in);
    while (infile.get(ch)) {
        cout << ch;
        mod[lenm] = ch;
        lenm ++;
    }
    infile.close();
```

```
        while (i < lenm) {
            while (j < lenk) {
                if (key[j] == mod[i]) {
                    i ++;
                    j ++;
                }
                else {
                    i = i - j + 1;
                    j = 0;
                    break;
                }
            }
            if (j == lenk)
            {
                number ++;
                j = 0;
            }
        }

    cout << '\n'<< "\n" ;
    cout << "在源文件中共检索到： " ;
    cout << number <<" 个关键字 \" ";
    cout << key << " \" "<< endl;
}


/***********************************************************************
    主函数
***********************************************************************/
void main()
{
    input();
    search();
}
```

3.3.1.4 运行结果

```
关键字检索系统
请输入文件名：src.txt

请输入一段英文：
With over two billion pages and more being added daily, the Web is a massive col
lection of interrelated pages. With so much available information, locating the
precise information you need can be difficult. Fortunately, a number of organiza
tions called search services or search providers can help you locate the informa
tion you need. They maintain huge databases relating to information provided on
the Web and the Internet. The information stored at the databases includes addre
sses, content descriptions or classifications, and keywords appearing on Web pag
es and other Internet informational resources.^
本段文本已保存在文本文件"src.txt"中。

请输入要检索的关键字：   Web
显示源文件" src.txt " ：

With over two billion pages and more being added daily, the Web is a massive col
lection of interrelated pages. With so much available information, locating the
precise information you need can be difficult. Fortunately, a number of organiza
tions called search services or search providers can help you locate the informa
tion you need. They maintain huge databases relating to information provided on
the Web and the Internet. The information stored at the databases includes addre
sses, content descriptions or classifications, and keywords appearing on Web pag
es and other Internet informational resources.^

在源文件中共检索到：  3 个关键字 " Web "
Press any key to continue
```

3.3.2 四元线性方程组求解

3.3.2.1 系统简介

输入四元线性方程组的系数和常数，求此四元线性方程组的解。

3.3.2.2 设计思路

线性方程主的解法主要就分为直接解法和迭代法直接解法。主要有列主元高斯消去法、列主元直接三角解法和追赶法。从代数的角度来看，前两者在本质上是一样的，只不过用高斯迭代法实现起来更快一些。

用函数实现过程中，迭代法是一种很重要的方法。它能充分利用系数矩阵的稀疏性，减少内存占用量，而且程序简单，但缺点是计算量大，同时还有收敛性需要讨论。迭代法主要包括雅可比迭代法，高斯—赛德尔迭代法，逐次超松弛迭代法。

○数据结构

用二维数组 a[i][j]存储四元线性方程组的系数和常数实现起来较为容易，也可用链表进行存储。

○程序设计

本实验主要采用列主元高斯消去法，其算法描述如下：
1. 将方程用增广矩阵[A|b]=（a[i][j]）（[n][n+1]）表示。
2. 矩阵消元，其中矩阵用二维数组进行存储，对 k=1, 2, …, n-1 进行消元：

（1）选主元，找 I[k]∈{k, k+1, …, n}使得|a[I[k]]|=max|a[I[k]]|（i 的下限为 k，上限为 n）；

（2）如果 a[I[k]]=0，则矩阵 A 奇异，程序结束；否则执行（3）；

（3）如果 I[k]≠k，则交换第 k 行与第 I[k]行对应元素的位置，a[k][j]与 a[I[k]][j]交换，j=k，…，n+1；

（4）消元，对 i=k+1，…，n 分别进行计算。

3.3.2.3 程序清单

```cpp
#include "iostream.h"
/************************************************************************
    求解四元线性方程组的解
************************************************************************/
void account(float a[4][5])
{
    for(int i = 0; i < 4; i ++)
    {
        if (a[0][0] == 0) {
            int m = 1;
            while (a[m][0] == 0 && m < 4) {
                m ++;
            }
            if (m == 4) {
                break;
            }
            else
                for(int n = 0; n < 5; n ++)
                {
                    float temp;
                    temp = a[m][n];
                    a[m][n] = a[0][n];
                    a[0][n] = temp;
                }
        }
        int j = 0;
        while (a[i][j] == 0) {
            j ++;
        }
        if (j < 4) {
            int t = j;
            for(j = 0; j < 4; j ++)
```

```cpp
                    a[i][j] = a[i][j] / a[i][t];

                for(int q = 0; q < 4; q ++)
                {
                    if(q == i)
                        continue;
                    else {
                        float p = a[q][t];
                        for(j = 0; j < 5; j ++)
                            a[q][j] = a[q][j] - p * a[i][j];
                    }
                }
            }
        }

    if (a[3][3]) {
        cout << "\n 此四元线性方程组的解是： " << '\n';
        for(int i = 0; i < 4; i ++)
            cout << 'X' << i + 1 << " = " << a[i][4] << '\n';
    }
    else {
        cout << "\n 此方程组无解或有无数解！ " << '\n';
    }
}

/********************************************************************
    主函数，输入四元线性方程组的系数和结果
********************************************************************/
void main()
{
    float a[4][5];
    cout << "请输入四元线性方程组的系数和常数： " << '\n';
    for(int i1 = 0; i1 < 4; i1 ++)
    {
        for(int j1 = 0; j1 < 5; j1 ++)
        {
            float m;
            cin >> m;
            a[i1][j1] = m;
        }
```

```cpp
        }

        cout << "\n 四元线性方程组的系数和常数为: "<< endl;
        for(int i2 = 0; i2 < 4; i2 ++)
        {
            for(int j2 = 0; j2 < 5; j2 ++)
            {
                cout << a[i2][j2] << '\t' ;
            }
            cout << '\n' ;
        }

        cout << "\n 四元线性方程组为: " << endl;
        for(int i = 0; i < 4; i ++)
        {
            for(int j = 0; j < 5; j ++)
            {
                switch(j) {
                case 0:
                case 1:
                case 2:
                    cout << a[i][j] << ' ' << 'X' << j + 1 << " + ";
                    break;
                case 3:
                    cout << a[i][j] << ' ' << 'X' << j + 1 << " = ";
                    break;
                default:
                    cout << a[i][j] << ' ' << '\n';
                }
            }
        }
        account(a);
}
```

3.3.2.4 运行结果

```
请输入四元线性方程组的系数和常数：
2 -1 0 0 1
-1 2 -1 0 2
0 -1 2 -1 -2
0 0 -1 2 -1

四元线性方程组的系数和常数为：
2       -1      0       0       1
-1      2       -1      0       2
0       -1      2       -1      -2
0       0       -1      2       -1

四元线性方程组为：
2 X1 + -1 X2 + 0 X3 + 0 X4 = 1
-1 X1 + 2 X2 + -1 X3 + 0 X4 = 2
0 X1 + -1 X2 + 2 X3 + -1 X4 = -2
0 X1 + 0 X2 + -1 X3 + 2 X4 = -1

此四元线性方程组的解是：
X1 = 5
X2 = 4
X3 = 1
X4 = 0
Press any key to continue_
```

3.4 巩固提高

输入一个单词，检索并输出该单词所在的行号、该行中出现的次数以及在该行中的相应位置。

○实现思路

（1）输入要检索的文本文件名，打开相应的文件；

（2）输入要检索统计的单词；

（3）行计数器置初值 0；

（4）while（不是文件结束）{

    读入一行到指定串中；

    求出串长度；

    行单词计数器清 0；

    调用模式匹配函数匹配单词定位、该行匹配单词计数；

    行号计数器加 1；

    If（行单词计数器!=0）输出行号、该行有匹配单词的个数以及相应的位置；

  }

第四章 树和二叉树

　　树型结构是一类重要的非线性数据结构。其中以树和二叉树最为常用，直观看来，树是以分支关系定义的层次结构。树结构在客观世界中广泛存在，如人类社会的族谱和各种社会组织机构都可以用树来形象表示。树在计算机领域中也得到广泛应用，如在编译程序中，可用树来表示源程序的语法结构。又如在数据库系统中，树型结构也是信息的重要组织形式之一。本章通过三个模拟项目来学习树及二叉树的存储结构及其各种操作，首先通过使用有关树的操作实现家谱管理，其次通过使用二叉树的操作实现表达式求值，最后通过哈夫曼树的应用实现图像压缩编码优化。

4.1 实践目的和要求

4.1.1 实践目的

1）掌握树和二叉树的基本概念和性质；
2）掌握树与二叉树的遍历方法及它们所确定的序列之间的关系；
3）掌握上机实现树和二叉树的基本方法；
4）掌握用指针类型描述、访问和处理二叉树的运算；
5）掌握二叉树的基本操作：建立二叉树、二叉树的遍历等运算在链式存储结构上的实现；
6）掌握二叉树的遍历思想，学会利用递归方法编写对二叉树这种递归数据结构进行处理的算法。

4.1.2 实践要求

1）认真阅读和掌握本章实践的程序。
2）上机运行本章的程序，保存和打印出程序的运行结果，并结合程序进行分析。
3）按照你对二叉树的操作需要，重新改写程序并运行，打印出文件清单和运行结果。
4）注意理解递归算法的执行步骤。
5）重点理解如何利用栈结构实现非递归算法。

4.2 基本概念

4.2.1 树

4.2.1.1 树的定义

树是由 n（n≥0）个结点（元素）组成的有限集合 T。其中，如果 n=0，则它是一棵空树；如果 n>0，则它满足以下两个条件：

1) 这 n 个结点中有且仅有一个结点作为树的根结点，简称为根结点；
2) 其余的结点可分成 m（m≥0）个互不相交的有限集 $T_1, \cdots, T_m$，其中每个集合本身又都是一棵符合本定义的树，称 $T_1, \cdots, T_m$ 为根结点的子树。

从这里可以看出树的定义是递归的，因为在树的定义中又用到树的定义，它刻画了树的固有特性，即一棵树由若干棵子树构成，子树又由更小的若干棵子树构成。所以树特别适合于表示元素之间的层次关系。

4.2.1.2 树的相关术语

结点的度：一个结点的子树数目。
树的度：树中各结点的度的最大值。
结点的层次：假设根结点的层次为 1，第 i 层的结点的子树根结点的层次为 i+1。
树的深/高度：树中叶子结点的最大层次。
叶子结点（终端结点）：度为 0 的结点。
分支结点（非终端结点）：度大于 0 的结点。
内部结点：除根结点外的分支结点。
结点的路径长度：从树中一个结点到另一个结点之间的分支构成的这两个结点之间的路径，路径上的分支数目就称为路径长度。
树的路径长度：从树根到每一结点的路径长度之和。
树的带权路径长度：树中所有叶子结点的带权路径长度之和。
孩子结点与双亲结点：树中结点的子树的根结点。相反，称该结点为孩子结点的双亲。
兄弟结点：具有同一个双亲的结点互称为兄弟结点。
有序树与无序树：如果将树中结点的各子树看成从左到右是有次序的（即不能交换），则称该树为有序树，否则为无序树。
森林：是 m（≥0）棵互不相交的树的集合。对树中每一个结点而言，其子树的集合即为森林。
满 m 次树：如果除根结点和叶子结点外，其他结点的度均为 m，且所有叶子结点均在同一层，这样的树称为满 m 次树。

4.2.1.3 树的性质

性质1：树中的结点数等于所有结点的度数加1。
性质2：度为 m 的树中第 i 层上至多有 m^{i-1} 个结点（i≥1）。
性质3：高度为 h 的 m 次树至多有 $(m^h-1)/(m-1)$ 个结点。

性质 4：具有 n 个结点的 m 次树的最小高度为 $\lceil \log_m(n(m-1)+1) \rceil$。

4.2.1.4 树的遍历

在应用树结构时，常要求按某种次序获得树中全部结点的信息，这可通过树的遍历操作来实现。

树的遍历是指按一定规律走遍树中的每一个结点，且使每一结点仅被访问一次，即找一个完整而有规律的走法，以得到树中所有结点的一个线性排列。

树的常用遍历方法有：

①先根遍历：先访问树的根结点，然后从左到右依次先根遍历根结点的每一棵子树。

②后根遍历：先从左到右依次后根遍历根结点的每一棵子树，然后访问根结点。

③按层次遍历：首先访问处于 0 层上的根结点，然后从左到右依次访问处于 1 层、2 层、……之上的结点，即自上而下从左到右逐层访问树各层上的结点。

4.2.1.5 树的存储结构

树是非线性的结构，不能简单地用结点的线性表来表示。树的主要存储结构有双亲存储结构、孩子链存储结构及孩子兄弟链存储结构三种。

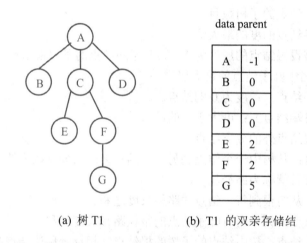

(a) 树 T1　　　(b) T1 的双亲存储结

图 4.1　树及其双亲存储结构

（1）双亲存储结构

这种存储结构是一种顺序存储结构，用一组连续空间存储树的所有结点，每个结点含两个域，数据域存放结点本身信息、双亲域指示本结点的双亲结点在数组中位置。

树及其双亲存储结构如图 4.1 所示。

该存储结构利用了除根结点外的每个结点只有惟一双亲的性质。

特点：求结点的双亲容易，但求结点的孩子结点困难，需要遍历整个结构。

（2）孩子链存储结构

由于树中每个结点的子树个数即结点的度不同，如果按各个结点的度设计变长结构，则每个结点的孩子结点指针域个数增加，使算法实现非常麻烦。孩子链存储结构可按树的度设计结点的孩子结点指针域个数，将每个结点的孩子结点用单链表存储，则 n 个结点有 n 个孩子链表，而 n 个头指针又组成了一个线性表。

树及其孩子链存储结构如图 4.2 所示。

特点：查找结点的双亲结点麻烦，需要从树的根结点开始逐个结点比较查找，浪费较多的指针空间，但适于存储某些特殊的树如二叉树、四叉树等。

（3）孩子兄弟链存储结构（又称二叉树存储结构、二叉链表存储结构）

孩子兄弟链存储结构是用二叉链表作树的存储结构，链表中每个结点设计三个域，数据域存放结点本身信息、两个指针域分别指向该结点的第一个孩子结点和该结点的下一个兄弟结点。树及其孩子兄弟链存储结构如图 4.3 所示。

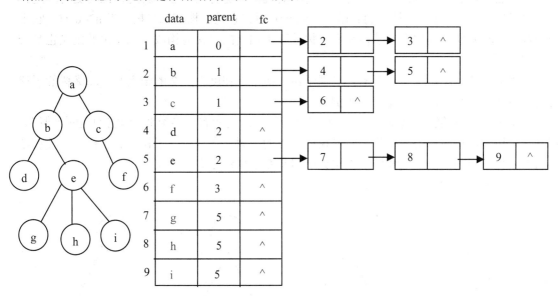

（a）树 T1　　　　　　　　　　（b）T1 的孩子链存储结构

图 4.2　树及其孩子链存储结构

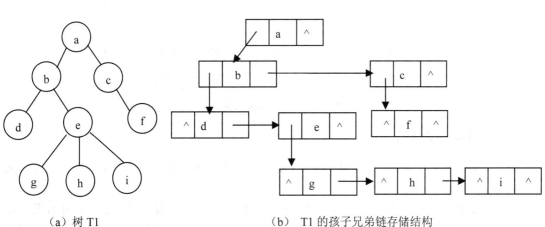

（a）树 T1　　　　　　　　　　（b）T1 的孩子兄弟链存储结构

图 4.3　树及其孩子兄弟链存储结构

由于树的孩子兄弟链存储结构有两个指针域，并且这两个指针是有序的，所以孩子兄弟链存储结构是把树转换为二叉树的存储结构。

特点：可方便地实现树和二叉树的相互转换。但查找结点的双亲结点麻烦，需要从树的根结点开始逐个结点比较查找而且破坏了树的层次。

4.2.2 二叉树

4.2.2.1 二叉树的定义

与一般的树的结构比较，二叉树在结构上更规范和更有确定性，应用也比树更为广泛。

二叉树是 n（n≥0）个结点的有限集，当 n=0 时，它是一棵空二叉树；当 n>0 时，它由一个根结点和两棵分别称为左子树和右子树的互不相交的二叉树构成。二叉树的定义也是一个递归定义。

二叉树的特点：每个结点至多有二棵子树（即不存在度大于 2 的结点）；二叉树的子树有左、右之分，且其次序不能任意颠倒。

二叉树不是树的特殊情形，尽管树和二叉树的概念之间有许多类似，但它们是两个概念。树和二叉树之间最主要的差别是：二叉树中结点的子树要区分为左子树和右子树，即使在结点只有一棵子树的情况下也要明确指出该子树是左子树还是右子树。

二叉树有五种基本形态，如图 4.4 所示。

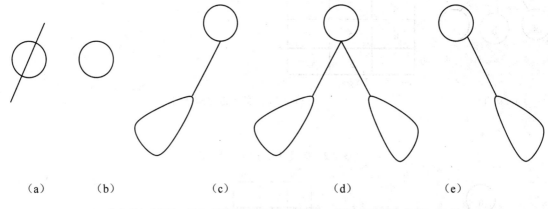

（a）　　　　（b）　　　　（c）　　　　　（d）　　　　　（e）

（a）空二叉树　（b）只有根结点的二叉树　（c）右子树为空的二叉树

（d）左、右子树均非空的二叉树　（e）左子树为空的二叉树

图 4.4　二叉树的基本形态

4.2.2.2 二叉树的相关术语

满二叉树：如果一棵二叉树的任何结点或者是叶子结点、或有两棵非空子树，则此二叉树称作满二叉树。

完全二叉树：如果一棵二叉树至多只有最下面的两层结点度数可以小于 2，并且最下面一层的结点都集中在该层最左边的若干位置上，则此二叉树称为完全二叉树。完全二叉树不一定是满二叉树。

4.2.2.3 二叉树的性质

性质 1：在二叉树的第 i 层上至多有 2^{i-1} 个结点（i≥1）。

性质2：深度为 k 的二叉树中至多有 2^k-1 个结点（k≥1）。

性质3：对任何一棵二叉树 T，若其有 n0 个叶子结点、n2 个度为 2 的结点，则 n0=n2+1。

性质4：具有 n（n＞0）个结点的完全二叉树的深度为 $\lceil \log_2(n+1) \rceil$ 或 $\lfloor \log_2 n \rfloor +1$。

性质5：若对含 n 个结点的完全二叉树从上到下且从左至右进行 1 至 n 的编号，则对完全二叉树中任意一个编号为 i 的结点：

①若 i=1，则该结点是二叉树的根，无双亲；否则，编号为 $\lfloor i/2 \rfloor$ 的结点为其双亲结点；

②若 2i＞n，则该结点无左孩子；否则，编号为 2i 的结点为其左孩子结点；

③若 2i+1＞n，则该结点无右孩子结点；否则，编号为 2i+1 的结点为其右孩子结点。

4.2.2.4 二叉树的遍历

树的所有遍历方法都适用于二叉树，常用的二叉树遍历方法有 3 种。

①先（根）序遍历：若二叉树为空树，则空操作；否则，先访问根结点，然后先序遍历左子树，最后先序遍历右子树。

②中（根）序遍历：若二叉树为空树，则空操作；否则，先中序遍历左子树，然后访问根结点，最后中序遍历右子树。

③后（根）序遍历：若二叉树为空树，则空操作；否则，先后序遍历左子树，然后后序遍历右子树，最后访问根结点。

以上三种遍历方法都是递归定义的。

4.2.2.5 哈夫曼树

哈夫曼树又称为最优树，是一类带权路径长度最短的树。

哈夫曼树就是一棵 n 个叶子结点的二叉树，所有叶子结点的带权路径之和最小。

○算法描述

①给定 n 个节点的集合，每个节点都带权值；

②选两个权值最小的节点构造一棵新的二叉树，新二叉树的根节点的权值就是两个子节点权值之和；

③从 n 个节点中删除刚才使用的两个节点，同时将新产生的二叉树根节点放在节点集合中。

④重复②、③步，直到只有一棵树为止。

4.2.2.6 二叉树的存储结构

（1）顺序存储结构

二叉树的树形存储结构就是用一组地址连续的存储单元来存放二叉树的数据元素。其中结点的存放次序是：对该树中每个结点进行编号，其编号从小到大的顺序就是结点存放在连续存储单元的先后次序。若把二叉树存储在一维数组中（下标从 1 开始），则该编号就是其下标值。

二叉树中各结点的编号与等高度的完全二叉树中对应位置上结点的编号相同。其编号过程是：首先把根结点的编号定为 1，然后按照层次从上到下、每层从左到右的顺序，对每一结点进行编号。当它是编号为 i 的双亲结点的左孩子结点时，则它的编号应为 2i；当它是编号为 i 的双亲结点的右孩子结点时，则它的编号应为 2i+1。

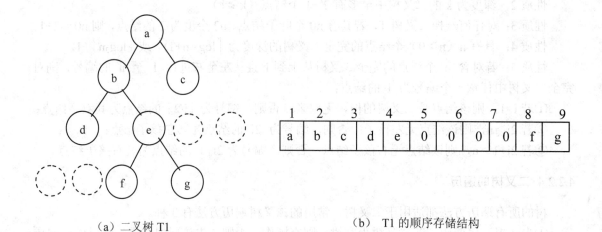

(a) 二叉树 T1　　　　　　　　　　(b) T1 的顺序存储结构

图 4.5　二叉树及其顺序存储结构

二叉树及其顺序存储结构如图 4.5 所示。

特点：由于结点间的关系蕴含在其存储位置中，因此访问每一个结点的双亲和左、右孩子结点（若存在的话）都非常方便。但是浪费空间，适于存储满二叉树和完全二叉树。

（2）二叉树的链式存储结构（二叉链存储结构）

二叉树中每一个结点用链表中的一个链结点来存储，每个结点有三个域：一个数据域用来存储结点本身的数据信息，另两个指针域分别指向其左、右孩子结点的位置。

二叉树及其链式存储结构如图 4.6 所示。

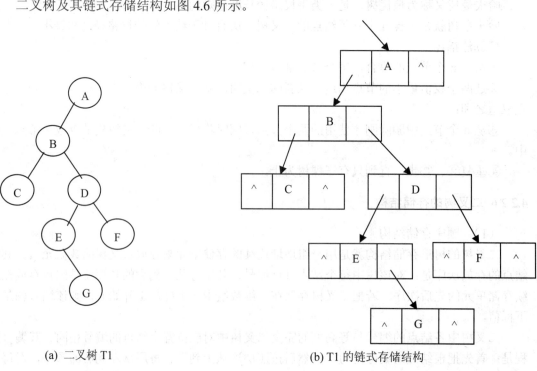

(a) 二叉树 T1　　　　　　　　(b) T1 的链式存储结构

图 4.6　二叉树及其链式存储结构

4.2.2.7 二叉树的基本运算

1）InitBiTree（&T）初始化：构造一棵空的二叉树 T。
2）CreateBiTree（&T，definition）建立：按 definition 的定义建立一棵二叉树 T。
3）BiTreeEmpty（T）判空：若二叉树 T 是为空则返回 TRUE，否则返回 FALSE。
4）Parent（T，e）求双亲：若 e 是 T 的非根结点，则返回它的双亲，否则返回 NULL。
5）LeftChild（T，e）求左孩子：若 e 是 T 的结点，则返回它的左孩子；若 e 无左孩子则返回 NULL。
6）RightChild（T，e）求右孩子：若 e 是 T 的结点，则返回它的右孩子；若 e 无右孩子则返回 NULL。
7）LeftSibling（T，e）求左兄弟：若 e 是 T 的结点，则返回它的左兄弟；若 e 无左兄弟则返回 NULL。
8）RightSibling（T，e）求右兄弟：若 e 是 T 的结点，则返回它的右兄弟；若 e 无右兄弟则返回 NULL。
9）PreOrderTraverse（T，Visit（））前序遍历：用 Visit（）前序遍历二叉树。
10）InOrderTraverse（T，Visit（））中序遍历：用 Visit（）中序遍历二叉树。
11）PostOrderTraverse（T，Visit（））后序遍历：用 Visit（）后序遍历二叉树。
12）LevelOrderTraverse（T，Visit（））层次遍历：用 Visit（）层次遍历二叉树。
13）BiTreeDepth（T）求深度：返回二叉树 T 的深度。
14）DestroyBiTree（&T）销毁：将二叉树 T 销毁。

4.3 实践案例

4.3.1 家谱管理系统

4.3.1.1 系统简介

家谱（或称族谱）是一种以表谱形式，记载一个以血缘关系为主体的家族世系繁衍和重要人物事迹的特殊图书体裁。家谱是中国特有的文化遗产，是中华民族的三大文献（国史，地志，族谱）之一，属珍贵的人文资料，对于历史学、民俗学、人口学、社会学和经济学的深入研究，均有其不可替代的独特功能。本项目对家谱管理进行简单的模拟，以实现查看祖先和子孙个人信息、插入家族成员、删除家族成员等功能。

4.3.1.2 设计思路

本项目的实质是完成对家谱成员信息的建立、查找、插入、修改、删除等功能，可以首先定义家族成员的数据结构，然后将每个功能作为一个成员函数来完成对数据的操作，最后完成主函数以验证各个函数功能并得出运行结果。

○数据结构

因为家族中的成员之间存在一个对多个的层次结构关系，所以不能用上面讲的线性表来表示。家谱从形状上看像一棵倒长的树，所以用树结构来表示家谱比较适合。树型结构是一类非常重要的非线性数据结构，直观看来，树是以分支关系定义的层次结构。

可以二叉链表作为树的存储结构，链表中的两个链域分别指向该结点的第一个孩子结点和下一个兄弟结点，因此该表示法又称二叉树表示法，或孩子兄弟表示法。

○程序设计和头文件定义

本系统采用孩子兄弟表示法链式存储结构存储家谱数据，其中节点类 Node，家谱成员类 Person 在头文件 Person.h 中的具体定义为：

```
/************************************************************************
    树的节点类
************************************************************************/
class Node
{
    friend class Person;
public:
    Node()
    {
        name = '?';
        lchild = NULL;
        rchild = NULL;
    }

private:
    string name;
    Node *lchild;
    Node *rchild;
};


/************************************************************************
    家谱成员类
************************************************************************/
class Person
{
public:
    void    Creat   ( Person &L );                      //创建家谱
    void    Add     ( Person &L );                      //添加部分家庭成员
    void    Update  ( Person &L );                      //家谱成员改名
    void    Insert  ( Person &L );                      //添加单个家庭成员
    void    Delete  ( Person &L );                      //删除部分家庭成员
    Node *  Lookup  ( Node *p, string name );           //查找某人
    void    Print   ( Node *p );                        //显示部分家庭成员
```

```cpp
    Person()
    {
        root = NULL;
    }

private:
    Node *root;
};
```

4.3.1.3 程序清单

```cpp
/*************************************************************************
    创建家谱
*************************************************************************/
void Person::Creat(Person &L)
{
    cout << "请输入祖先的姓名：";
    string rootname;
    cin >> rootname;
    Node *p = new Node;
    p->name = rootname;
    L.root = p;
    cout << "此家谱的祖先是：" << p->name << '\n';
}

/*************************************************************************
    添加部分家庭成员
*************************************************************************/
void Person::Add(Person &L)
{
    cout << "请输入要建立家庭的人的姓名：";
    string rootname;
    cin >> rootname;
    Node *s = Person::Lookup(L.root, rootname);
    if (s) {
        Node *r = s;
        cout << "请输入" << s->name << "的儿女人数：";
        int n;
        cin >> n;
        int m = n;
```

```cpp
            cout << "请依次输入" << s->name << "的儿女的姓名：";
            while (m) {
                Node *q = new Node;
                string name;
                cin >> name;
                q->name = name;
                if (m == n) {
                    s->lchild = q;
                    s = s->lchild;
                }
                else {
                    s->rchild = q;
                    s = s->rchild;
                }
                m --;
            }
            Person::Print(r);
        }
        else {
            cout << "查无此人，请重新输入！" << '\n';
            Person::Add(L);
        }
    }
}


/************************************************************************
    查找某人
************************************************************************/
Node * Person::Lookup(Node *p, string name)
{
    Node *t = NULL;
    Node *s[100];
    int top = 0;
    while (p || top > 0) {
        while (p) {
            if (p->name == name) {
                t = p;
            }
            s[++ top] = p;
            p = p->lchild;
        }
```

```cpp
            p = s[top --];
            p = p->rchild;
        }
        return t;
}


/*************************************************************************
    显示部分家庭成员
*************************************************************************/
void Person::Print(Node *p)
{
    cout << p->name << "的第一代子孙是：" << p->lchild->name << '\t';
    p = p->lchild;
    while (p->rchild) {
        cout << p->rchild->name << '\t';
        p = p->rchild;
    }
    cout << '\n';
}


/*************************************************************************
    添加单个家庭成员
*************************************************************************/
void Person::Insert(Person &L)
{
    cout << "请输入要添加儿子（或女儿）的人的姓名：";
    string rootname;
    cin >> rootname;
    Node *s = Person::Lookup(L.root, rootname);
    if (s) {
        Node *r = s;
        cout << "请输入" << s->name << "新添加的儿子（或女儿）的姓名：";
        Node *p = new Node;
        string name;
        cin >> name;
        p->name = name;
        if (!s->lchild) {
            s->lchild = p;
        }
        else {
```

```cpp
                s = s->lchild;
                while (s->rchild) {
                    s = s->rchild;
                }
                s->rchild = p;
            }
            Person::Print(r);
        }
        else {
            cout << "查无此人，请重新输入！" << '\n';
            Person::Insert(L);
        }
    }
}

/*******************************************************************
    删除部分家庭成员
*******************************************************************/
void Person::Delete(Person &L)
{
    cout << "请输入要解散家庭的人的姓名：";
    string rootname;
    cin >> rootname;
    Node *s = Person::Lookup(L.root, rootname);
    if (s) {
        if (s->lchild) {
            cout << "要解散家庭的人是：" << s->name << '\n';
            Person::Print(s);
            s->lchild = NULL;
        }
        else {
            cout << s->name << "尚未有家庭！";
        }
    }
    else {
        cout << "查无此人，请重新输入！" << '\n';
        Person::Delete(L);
    }
}
```

/***
    家谱成员改名
**/
```cpp
void Person::Update(Person &L)
{
    cout << "请输入要更改姓名的人的目前姓名：";
    string rootname;
    cin >> rootname;
    Node *s = Person::Lookup(L.root, rootname);
    if (s) {
        cout << "请输入更改后的姓名：";
        string name;
        cin >> name;
        s->name = name;
        cout << rootname << "已更名为" << s->name << '\n';
    }
    else {
        cout << "查无此人，请重新输入！" << '\n';
        Person::Update(L);
    }
}
```

/***
    主函数
**/
```cpp
void main()
{
    cout<<"\n**                  家谱管理系统                    **"<<endl;
    cout<<"======================================================"<<endl;
    cout<<"**            请选择要执行的操作 ：                 **"<<endl;
    cout<<"**                A --- 完善家谱                    **"<<endl;
    cout<<"**                B --- 添加家庭成员                **"<<endl;
    cout<<"**                C --- 解散局部家庭                **"<<endl;
    cout<<"**                D --- 更改家庭成员姓名            **"<<endl;
    cout<<"**                E --- 退出程序                    **"<<endl;
    cout<<"======================================================"<<endl;

    cout << "首先建立一个家谱！" << '\n';
    Person L;
    L.Creat(L);
```

```cpp
    char ch;
    while(ch!='E')
    {
        cout << "\n 请选择要执行的操作：";
        cin >> ch;
        switch(ch) {
        case 'A':
            {
                L.Add(L);
                break;
            }
        case 'B':
            {
                L.Insert(L);
                break;
            }
        case 'C':
            {
                L.Delete(L);
                break;
            }
        case 'D':
            {
                L.Update(L);
                break;
            }
        case 'E':
            break;
        default:
            cout << "请输入正确的操作！" << '\n';
        }
    }
}
```

4.3.1.4 运行结果

```
**        家谱管理系统         **
====================================
          请选择要执行的操作:
          A ---- 完善家谱           **
**        B ---- 添加家庭成员        **
**        C ---- 解散局部家庭        **
**        D ---- 更改家庭成员姓名     **
          E ---- 退出程序
====================================
首先建立一个家谱!
请输入祖先的姓名: P0
此家谱的祖先是: P0

请选择要执行的操作: A
请输入要建立家庭的人的姓名: P0
请输入P0的儿女人数: 2
请依次输入P0的儿女的姓名: P1 P2
P0的第一代子孙是: P1    P2

请选择要执行的操作: A
请输入要建立家庭的人的姓名: P1
请输入P1的儿女人数: 3
请依次输入P1的儿女的姓名: P11 P12 P13
P1的第一代子孙是: P11    P12    P13

请选择要执行的操作: B
请输入要添加儿子(或女儿)的人的姓名: P2
请输入P2新添加的儿子(或女儿)的姓名: P21
P2的第一代子孙是: P21

请选择要执行的操作: C
请输入要解散家庭的人的姓名: P2
要解散家庭的人是: P2
P2的第一代子孙是: P21

请选择要执行的操作: D
请输入要更改姓名的人的目前姓名: P13
请输入更改后的姓名: P14
P13已更名为P14

请选择要执行的操作: E
Press any key to continue_
```

4.3.2 表达式求值问题

4.3.2.1 系统简介

表达式求值是程序设计语言编译中的一个最基本问题,就是将一个表达式转化为逆波兰表达式并求值。具体要求是以字符序列的形式从终端输入语法正确的、不含变量的整数表达式,并利用给定的优先关系实现对算术四则混合表达式的求值,并演示在求值过程中运算符栈、操作数栈、输入字符和主要操作变化过程。

要把一个表达式翻译成正确求值的一个机器指令序列,或者直接对表达式求值,首先要能正确解释表达式。任何一个表达式都是由操作符(operand)、运算符(operator)和界限符(delimiter)组成,我们称它们为单词。一般来说,操作数既可以是常数,也可以是被说明为变量或常量的标识符;运算符可以分为算术运算符、关系运算符和逻辑运算符3类;基本界限符有左右括号和表达式结束符等。为了叙述的简洁,我们仅仅讨论简单算术表达式的求值问题。这种表达式只包括加、减、乘、除 4 种运算符。

人们在书写表达式时通常采用的是"中缀"表达形式,也就是将运算符放在两个操作数中间,用这种"中缀"形式表示的表达式称为中缀表达式。但是,这种表达式表示形式对计算机处理来说是不大合适的。对于表达式的表示还有另一种形式,称之为"后缀表达式",也就是将运算符紧跟在两个操作数的后面。这种表达式比较适合计算机的处理方式,因此要用计算机来处理、计算表达式的问题,首先要将中缀表达式转化成后缀表达式,又称为逆波兰表达式。

4.3.2.2 设计思路

○ **数据结构**

为了实现表达式求值,可以首先读入原表达式(包括括号)并创建对应二叉树,其次对二叉树进行前序遍历、中序遍历、后续遍历,输出对应的逆波兰表达式、中序表达式和波兰表达式。

○ **程序设计**

用二叉树的结构来存储表达式,首先以中序生成二叉树,先序遍历二叉树可得到逆波兰表达式,中序遍历二叉树可得到中缀表达式,后序遍历二叉树可得到波兰表达式。其中二叉树结点类型 Node 和二叉树类 Tree 在头文件 Tree.h 中的具体定义为:

```
/************************************************************************
    二叉树结点类
*************************************************************************/
class Node
{
    friend class Tree;
public:
    Node()
    {
        lchild = NULL;
        key = '\0';
        rchild = NULL;
    }

private:
    Node    * lchild;
    char      key;
    Node    * rchild;
};
/************************************************************************
    二叉树类
*************************************************************************/
```

```
class Tree
{
public:
    void    rootLast    ( Node   *p );           //先序遍历二叉树
    void    rootMiddle  ( Node   *p );           //中序遍历二叉树
    void    rootFirst   ( Node   *p );           //后序遍历二叉树
    void    Creat       ( Tree &L,  string exp );  //以中序生成二叉树
    Tree()
    {
        root = NULL;
    }
    Node *root;
};
```

4.3.2.3 程序清单

```
/***********************************************************************
    以中序生成二叉树
***********************************************************************/
void Tree::Creat(Tree &L, string exp)
{
    Node *s = new Node;
    s->key = exp[0];
    L.root = s;
    int i = 1;
    while (exp[i] != '\0') {
        Node *p = new Node;
        p->key = exp[i];
        if (i % 2 == 1) {
            p->lchild = L.root;
            L.root = p;
        }
        else {
            L.root->rchild = p;
        }
        i ++;
    }
}
```

/**
    后序遍历二叉树，输出波兰表达式
**/
void Tree::rootFirst(Node *p)
{
    if(p!=NULL)
    {
        cout << p->key;
        rootFirst(p->lchild);
        rootFirst(p->rchild);
    }
}

/**
    中序遍历二叉树，输出中缀表达式
**/
void Tree::rootMiddle(Node *p)
{
    if(p!=NULL)
    {
        rootMiddle(p->lchild);
        cout << p->key;
        rootMiddle(p->rchild);
    }
}

/**
    先序遍历二叉树，输出逆波兰表达式
**/
void Tree::rootLast(Node *p)
{
    if(p!=NULL)
    {
        rootLast(p->lchild);
        rootLast(p->rchild);
        cout << p->key;
    }
}

```
/***********************************************************************
    主函数
***********************************************************************/
void main()
{
    cout << "请输入表达式：\t";
    string exp;
    cin >> exp;
    cout << '\n';

    Tree L;
    L.Creat(L, exp);

    cout << "波兰表达式：\t";
    L.rootFirst(L.root);
    cout << '\n' << '\n';

    cout << "中缀表达式：\t";
    L.rootMiddle(L.root);
    cout << '\n' << '\n';

    cout << "逆波兰表达式：\t";
    L.rootLast(L.root);
    cout << '\n' << '\n';
}
```

4.3.2.4 运行结果

```
请输入表达式：    5+3*(7-8/2)+6
波兰表达式：      +287*+53(-/)6
中缀表达式：      5+3*(7-8/2)+6
逆波兰表达式：    53+(*-7/8)26+
Press any key to continue
```

4.3.3 图像压缩编码优化问题

4.3.3.1 系统简介

信息时代，人们对使用计算机获取信息、处理信息的依赖性越来越高。计算机系统面临的是数值、文字、语言、音乐、图形、动画、图像、电视视频图像等多种媒体。数字化的视频和音频信号的数量之大是惊人的，对于电视画面的分辨率 640×480 的彩色图像，30 帧/s，则一秒钟的数据量为：640×480×24×30＝221.12M，所以播放时，需要221Mbps的通信回路。存储时，1张CD可存640M，仅可以存放2.89s的数据。多媒体信息中，视频信息是一种比较特殊的媒体，数据量极大，信息丰富，并以与时间密切相关的流的形式存在。因此，视频数据的表达、组织、存储和传输都有很大难度。解决的基础在于对视频数据进行压缩。自1948年Oliver提出PCM编码理论以来，编码压缩技术日趋成熟，产生了许多压缩算法和技术，如在H.261、JPEG和MPEG编码标准中的应用等。数据压缩目前的主要目标是较大的压缩比、较快的压缩解压速度以及尽可能好的图像还原质量，而对于压缩数据的处理如数据组织、检索、重构等还并没有较好的考虑，也没有一个比较完整的解决方案，因此在这方面仍有许多工作要作。最基本的是要有更加合理高效的压缩算法。

大量的图像信息数据会给存储器的存储容量，通信干线信道的带宽，以及计算机的处理速度增加极大的压力。单纯靠增加存储器容量，提高信道带宽以及计算机的处理速度等硬件方法来解决这个问题是不现实的，这时就要考虑软件压缩方法。压缩的关键在于编码，如果在对数据进行编码时，对于常见的数据，编码器输出较短的码字，而对于少见的数据则用较长的码字表示，就能够实现数据压缩。

4.3.3.2 设计思路

在对一幅大小为100 672bytes 8位BMP图像文件进行Huffman编码过程中，按照以下步骤对图像进行压缩和解压缩：

1）扫描位图文件的全部数据（对应用于调色板的编码），完成数据频度的统计。
2）依据数据出现的频度建立哈夫曼树。
3）将哈夫曼树的信息写入输出文件（压缩后文件），以备解压缩时使用。
4）进行第二遍扫描，将原文件所有编码数据转化为哈夫曼编码，保存到输出文件。解压缩则为逆过程。

○**数据结构**

假设一个数据文件中出现了8种符号S0，S1，S2，S3，S4，S5，S6，S7，那么每种符号要编码，至少需要3bit。假设S0到S7的编码是000，001，010，011，100，101，110，111。那么文件中的数据符号序列 S0S1S7S0S1S6S2S2S3S4S5S0S0S1 编码后变成000001111000001110010010011100101000000001，共用了42bit。

我们发现S0，S1，S2这3个符号出现的频率比较大，其他符号出现的频率比较小，我们采用这样的编码方案：S0到S7的码字分别是01，11，101，0000，0001，0010，0011，100，那么上述符号序列变成01111000111001110110100000010010010111，共用了39bit。尽管有些码字如S3，S4，S5，S6变长了（由3位变成4位），但使用频繁的几个码字如S0，S1变短了，所以实现了压缩。

对于上述的编码可能导致解码出现非单值性：比如说，如果S0的码字为01，S2的码字为011，那么当序列中出现011时，你不知道是S0的码字后面跟了个1，还是完整的一个S2的码字。因此，编码必须保证较短的编码决不能是较长编码的前缀。符合这种要求的编码称之为前缀编码。

要构造符合这样的二进制编码体系，可以通过二叉树来实现。

1）首先统计出每个符号出现的频率，上例S0到S7的出现频率分别为4/14，3/14，2/14，1/14，1/14，1/14，1/14，1/14。

2）从左到右把上述频率按从小到大的顺序排列。

3）每一次选出最小的两个值，作为二叉树的两个叶子节点，将和作为它们的根节点，这两个叶子节点不再参与比较，新的根节点参与比较。

4）重复3），直到最后得到和为1的根节点。

将形成的二叉树的左节点标 0，右节点标 1。把从最上面的根节点到最下面的叶子节点路径中遇到的 0，1 序列串起来，得到各个符号的编码。

可以看到，符号只能出现在树叶上，任何一个字符的路径都不会是另一字符路径的前缀路径，这样，前缀编码也就构造成功了。

这样一棵二叉树称之为 Huffman 树，常用于最佳判定，它是最优二叉树，是一种带权路径长度最短的二叉树。所谓树的带权路径长度，就是树中所有的叶结点的权值乘上其到根结点的路径长度（若根结点为 0 层，叶结点到根结点的路径长度为叶结点的层数）。树的带权路径长度记为：WPL＝（W1*L1＋W2*L2＋W3*L3＋…＋Wn*Ln），N 个权值 Wi（i＝1, 2，…，n）构成一棵有 N 个叶结点的二叉树，相应的叶结点的路径长度为 Li（i＝1, 2，…，n）。Huffman 树得出的 WPL 值最小。

○ 程序设计

Huffman 树结点类的数据结构应有 511 个节点，这是因为每个字节可表示的符号个数为 2^8＝256 个（对应于 256 种颜色）。二叉树有 256 个叶节点，根据二叉树的性质总节点数为 2·256－1＝511 个节点。这里用 0～255 个元素来依次对应 256 种颜色。由第 256 以后的元素来依次对应形成的各个父节点的信息，即父节点的编号从 256 开始。

按照前述的压缩步骤，先对欲压缩文件的各个符号的使用次数进行统计，填充于 HuffNode［0～255］·freq 项内；在已有的节点中找出频率最低的两个节点，给出它们的父节点，将两个节点号填充于父节点的 lchild，rchild，将父节点号填充于两个节点的 parent 内。重复步骤直到根节点，建树工作完成。

建树完成后进行编码，对每个符号从符号的父节点开始。若节点的父节点值不为－1，则一直进行下去，直到树根。回溯过程中遇左出 0、遇右出 1 输出编码。解码时从树根开始，遇 1 取右节点，遇 0 取左节点，直到找到节点号小于 256 的节点（叶节点）。

○ 头文件定义

Huffman 树结点类在头文件 Huffman.h 中的具体定义为：

```
/************************************************************************
    Huffman 树结点类
*************************************************************************/
class HuffNode
{
    friend class HuffTree;
public:
    HuffNode();
    ~HuffNode();


private:
    int power;
```

```
    char key;
    int freq;
    HuffNode * lchild;
    HuffNode * rchild;
    HuffNode * parent;
};
```

Huffman 树类在头文件 Huffman.h 中的具体定义为:

```
/***************************************************************************
    Huffman 树类
***************************************************************************/
class HuffTree
{
public:
    void Print(HuffNode *t);
    void ShowHuff(HuffTree &L, char *key, int n);
    void BuildHuffman(HuffTree &L, int *freq, char *key, int n);

    HuffTree();
    ~HuffTree();

private:
    HuffNode * root;
};
```

4.3.3.3 程序清单

```
const int MAXV = 100;

/***************************************************************************
    Huffman 树结点构造函数
***************************************************************************/
HuffNode::HuffNode()
{
    lchild = NULL;
    power = MAXV;
    key = '\0';
    freq = 0;
    rchild = NULL;
    parent = NULL;
}
```

/***
    Huffman 树结点析构函数
***/
HuffNode::~HuffNode()
{
    delete lchild;
    delete rchild;
    delete parent;
}

/***
    Huffman 树类构造函数
***/
HuffTree::HuffTree()
{
    root = new HuffNode();
}

/***
    Huffman 树类析构函数
***/
HuffTree::~HuffTree()
{
    delete root;
}

/***
    构造 Huffman 树
***/
void HuffTree::BuildHuffman(HuffTree &L, int *freq, char *key, int n)
{
    HuffTree *a = & L;

    for(int i = 0; i < n; i ++)
    {
        HuffNode *p = new HuffNode();
        p->freq = freq[i];
        p->key = key[i];
        a[i].root = p;
    }

```cpp
    for(i = n; i < (2 * n - 1); i ++)
    {
        HuffNode *p = new HuffNode;
        p->freq = 0;
        p->key = '\0' ;
        a[i].root = p;
    }
    for(i = n; i< (2 * n - 1); i ++)                    //建哈夫曼树
    {
        int t = 0, d = 0;
        for(int j = 1; j < i; j ++)
        {
            if(a[t].root->freq >= a[j].root->freq && a[j].root->freq < MAXV)
            {
                d = t;
                t = j;
            }
        }
        if (t == 0) {
            d = 1;
            for(j = 2; j < i; j ++)
            {
                if(a[d].root->freq >= a[j].root->freq && a[j].root->freq < MAXV)
                {
                    d = j;
                }
            }
        }
        //将找出的两棵权值最小的子树合并为一棵子树
        a[i].root->freq = a[d].root->freq + a[t].root->freq;
        a[i].root->lchild = a[d].root;
        a[i].root->rchild = a[t].root;
        a[d].root->freq = MAXV;
        a[d].root->power = 0;
        a[d].root->parent = a[i].root;
        a[t].root->freq = MAXV;
        a[t].root->power = 1;
        a[t].root->parent = a[i].root;
    }
```

```cpp
        L.root = a[i-1].root;

}

/************************************************************************
        显示 Huffman 编码，并调用 Print()显示 Huffman 树
************************************************************************/
void HuffTree::ShowHuff(HuffTree &L, char *key, int n)
{
        cout << "\nHuffman 编码： " << endl;
        HuffNode *p , *t;
        HuffNode *s[100];

        for (int i = 0; i < n; i ++)
        {
                p = L.root ;
                int top=0;
                while (p || top > 0) {
                        while (p) {
                                if (p->key == key[i]) {
                                        t = p;
                                }
                                s[++ top] = p;
                                p = p->lchild;
                        }
                        p = s[top --];
                        p = p->rchild;
                }
                cout << key[i] << "的 Huffman 编码是： ";
                HuffTree::Print(t);
                cout << '\n';
        }
        cout << endl;
        delete p;
        delete t;
        delete[] *s;
}
/************************************************************************
        输出 Huffman 树
************************************************************************/
void HuffTree::Print(HuffNode *t)
```

```cpp
{
    if( t->parent )
    {
        HuffTree::Print(t->parent);
        cout << t->power;
    }
}
/*************************************************************************
    主函数
*************************************************************************/
void main()
{
    cout << "请输入字符个数：";
    int n;
    cin >> n;
    cout << "请依次输入字符和频率：" << '\n';
    char *key = new char[];
    int *freq = new int [];
    for(int i = 0; i < n; i ++)
    {
        char name;
        cin >> name;
        key[i] = name;
        int m;
        cin >> m;
        freq[i] = m;
    }
    key[i+1]='\0';

    HuffTree L ;
    HuffTree *a = new HuffTree [];
    a[0] = L ;

    L.BuildHuffman(L, freq, key, n);
    L.ShowHuff(L, key, n);

    delete[] key;
    delete[] freq;
    delete[] a;
}
```

4.3.3.4 运行结果

```
请输入字符个数：3
请依次输入字符和频率：
a 3
b 2
c 1

Huffman编码：
a的Huffman编码是：0
b的Huffman编码是：10
c的Huffman编码是：11
Press any key to continue_
```

4.4 巩固提高

表达式求值时可解析变量，同时对非法表达式格式能予以判断，最后可以通过树的合并操作完成表达式的计算。

○实现思路

设 str1=（a+b）*（(c+d)*e+f*h*g）；str2 是目的串。

先用一个堆栈存储 operand，遇到数字就不作处理直接放到目的串中，遇到 operaor 则将其与堆栈顶元素比较优先级，决定是否送到输出串中，这样处理之后能得到一个后续的排列，再建立二叉树，当遇到 ab+这样的建立左右接点为 a 和 b 的根为+的树并返回根接点，当遇到操作符前面只有一个操作数的如 c*这样的情况就建立右子接点为 c 的根接点为*的，左结点为上次操作返回的子树的根接点（这种情况下就是合并树的情况）。

非递归方法进行树的遍历需要用到栈结构。

首先调用 stack 的 ADT 中的 creatEmptyStack（）函数来建立一个空栈（注意，栈元素的类型是指向树结点的指针）。

再将根节点压入栈中，然后进入循环；如果栈不空，则弹出栈顶元素并对其进行访问，然后循环压入右子树（节点指针）和左子树（节点指针）。

○提示

由于需要同时用到树和栈两种抽象数据类型并且需要进行互动操作，在头文件中需要加入#ifndef#define#endif 宏命令来避免出现重复定义的问题。

第五章 图

图是一种较线性表和树更为复杂的数据结构，也是日常生活中应用十分广泛的数据结构之一。在线性表中，数据元素之间仅仅有线性关系，每个数据元素只有一个直接前驱和一个直接后继。在树形结构中，数据元素之间有着明显的层次关系，并且每一层上的数据元素可能和下一层中多个元素（即其孩子结点）相关，但只能和上一层中的一个元素（即其双亲结点相关）。而在图形结构中，结点之间的关系可以是任意的，图中任意两个数据元素之间都可能相关。由此，图的应用极为广泛，特别是近年来的迅速发展，已渗入到诸如语言学、逻辑学、物理、化学、电信工程、计算机科学以及数学的许多分支中。本章通过公交线路管理模拟系统、导航最短路径查询系统、电网建设造价系统和软件工程进程规划系统四个项目来学习图的定义、表示方法及其实现。

5.1 实践目的和要求

5.1.1 实践目的

1) 掌握图的概念、图的两种存储结构（邻接矩阵和邻接表）的存储思想及其存储实现；
2) 掌握上机实现图的基本方法；
3) 掌握有关图的操作并用高级语言编程实现；
4) 熟练掌握图的深度、广度优先遍历算法思想及其程序实现；
5) 掌握图的常见应用算法的思想及其程序实现。

5.1.2 实践要求

1) 掌握本章实践的算法。
2) 上机运行本章的程序，保存和打印出程序的运行结果，并结合程序进行分析。
3) 按照你对图的操作需要，重新改写程序并运行，打印出文件清单和运行结果。
4) 注意理解各算法实现时所采用的存储结构。
5) 注意区别正、逆邻接表。

5.2 基本概念

5.2.1 图的定义

图是一种较线性表和树更为复杂的数据结构。在图形结构中，结点之间的关系可以是任意的，图中任意两个数据元素之间都可能相关。

图：图 G 由两个集合 V 和 E 组成，记为 G=（V，E），其中 V 是顶点的有穷非空集合，

E 是 V 中顶点偶对（称为边）的有穷集。通常，也将图 G 的顶点集和边集分别记为 V（G）和 E（G）。E（G）可以是空集，若 E（G）为空，则图 G 只有顶点而没有边，称为空图。

5.2.2 图的相关术语

无向图：在图 G 中，如果代表边的顶点偶对是无序的，则称 G 为无向图。用 (i, j) 来表示顶点 i 和 j 之间的边。

有向图：在图 G 中，如果代表边的顶点偶对是有序的，则称 G 为有向图。有向图的边也称为弧，用 <i, j> 来表示顶点 i 和 j 之间的弧。

完全图：若图 G 中的每两个顶点之间都存在一条边，则我们称 G 为完全图。完全无向图有 n(n-1)/2 条边，完全有向图有 n(n-1) 条边。

端点和邻接点：在一个无向图中，若存在一条边 (v_i, v_j)，则称顶点 v_i 和 v_j 为此边的两个端点，也称它们互为邻接点，这里，v_i 为起点，v_j 为终点。

顶点的度、入度和出度：在无向图中，顶点 v 的度是关联于该顶点的边的数目，通常记为 D(v)。在有向图中，把以顶点 v 为终点的入边的数目称为 v 的入度，记为 ID(v)；把以顶点 v 为起点的出边的数目称为 v 的出度，记为 OD(v)；有向图中顶点 v 的度则定义为该顶点的入度和出度之和，即 D(v)=ID(v)+OD(v)。

子图：如果有两个图 G（V, E）和图 G'（V', E'），满足：V'⊆V 和 E'⊆E，则称 G' 为 G 的子图。

路径：在一个图 G=（V, E）中，路径是指从顶点 v_i 到顶点 v_j 所经过的所有顶点的序列 $(v_i, v_{i_1}, v_{i_2}, \cdots, v_{i_m}, v_j)$。若 G 是无向图，则边 (v_i, v_{i_1})，(v_{i_1}, v_{i_2})，…，$(v_{i_{m-1}}, v_{i_m})$，(v_{i_m}, v_j) 均属于 E（G）；若 G 是有向图，则边 $<v_i, v_{i_1}>$，$<v_{i_1}, v_{i_2}>$，…，$<v_{i_{m-1}}, v_{i_m}>$，$<v_{i_m}, v_j>$ 均属于 E（G）。路径长度定义为一条路径上经过的边的数目。若一条路径上除起始点和结束点可以相同外，其余顶点均不重复出现的路径叫简单路径。

回路或环：如果一条路径上的起始点和结束点相同，则这条路径称为回路或环。起始点与结束点相同的简单路径称为简单回路或简单环。

连通、连通图和连通分量：在无向图中，若从顶点 v_i 到顶点 v_j 有路径，则称 v_i 和 v_j 是连通的。如果图中任意两个顶点都连通，则该无向图称为连通图，否则称为非连通图。无向图中的极大连通子图称为该图的连通分量。显然，任何连通图的连通分量只有一个即本身，而非连通图可能有多个连通分量。

强连通图和强连通分量：在有向图中，若从顶点 v_i 到顶点 v_j 有路径，则称从 v_i 到 v_j 是连通的。如果图中任意两个顶点都连通，即每一对顶点 v_i, $v_j \in V$, $v_i \neq v_j$，从 v_i 到 v_j 和从 v_j 到 v_i 都存在路径，则称 G 是强连通图，否则称为非强连通图。有向图中的极大强连通强子图称为该图的强连通分量。显然，强连通图的连通分量只有一个即本身，而非强连通图可能有多个强连通分量。

稠密图、稀疏图：当一个图接近完全图时称为稠密图。反之，当一个图含有较少的边数即 e << n(n-1) 时称为稀疏图。

权和网：图的每一条边或弧都可以具有与它相关的数，这种与图的边或弧相关的数称为权。边或弧上带权的图称为权图，也称作网。

5.2.3 图的存储结构

图有多种存储方式，其中最常用的两种存储方式是邻接矩阵和邻接表。

5.2.3.1 邻接矩阵（数组表示法）

邻接矩阵是反映顶点间邻接关系的矩阵。定义如下：设 G=（V，E）是具有 n（n≥1）个顶点的图，G 的邻接矩阵 M 是一个 n 行 n 列的矩阵，若（i, j）或<i, j>∈E，则 M[i][j]=1；否则 M[i][j]=0。若是带权图，图中顶点 vi 与顶点 vj 邻接且权值为 wi，则邻接矩阵的 M[i][j]=wi，否则 M[i][j]=∞。图及其邻接矩阵如图 5.1 所示。

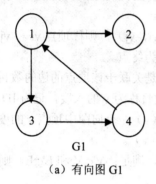
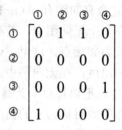

（a）有向图 G1　　　　　　　　　　（b）G1 的邻接矩阵

图 5.1　图及其邻接矩阵

在这种表示法中需要用两个数组分别存放图中顶点的信息（数据元素）和图中边（或弧）的信息（数据元素之间的关系）。

○邻接矩阵表示法的特点

①无向图的邻接矩阵对称，可压缩存储；有 n 个顶点的无向图需存储空间为 n（n+1）/2；

②有向图的邻接矩阵不一定对称；有 n 个顶点的有向图需存储空间为 n^2；

③无向图的邻接矩阵的第 i 行（或第 i 列）非零元素（或非∞元素）个数为第 i 个顶点的度 D（vi）；有向图的邻接矩阵的第 i 行非零元素（或非∞元素）个数为第 i 个顶点的出度 OD（vi），第 i 列非零元素（或非∞元素）个数就是第 i 个顶点的入度 ID（vi）；

④邻接矩阵表示图，很容易确定图中任意两个顶点之间是否有边相连；

⑤图的邻接矩阵表示是惟一的。

○邻接矩阵表示法的优点：各种基本操作都易于实现。

○邻接矩阵表示法的缺点：空间浪费严重，某些算法时间效率低（如图的创建）。

5.2.3.2 邻接表

邻接表是图的链式存储结构。邻接表表示法就是对图中的所有顶点都建立一个单链表来存储所有与该顶点邻接的顶点。每个单链表的第一个结点存放有关顶点的信息，把这一结点看成链表的表头，其余结点存放有关边的信息。因此邻接表是由单链表的表头形成的顶点表和单链表其余结点形成的边表两部分组成。一般顶点表存放顶点信息以及指向该顶点的第一个邻接点的指针。边表存放被邻接顶点的位置和指向下一个邻接点的指针。

图的邻接表如图 5.2（b）所示。

从邻接表的存储结构可以看出，对于有 n 个顶点和 e 条边的无向图，它的邻接表需要 n

个头结点的空间和 2e 条边的空间,每个单链表的结点数就是相应顶点的度。有向图边表中的结点数等于图中边数,每个单链表的结点数就是相应顶点的出度。而顶点 i 的入度为整个邻接表中邻接点域值是 i 的结点个数。

对于有向图来说,由于求邻接表中某个顶点的入度不方便,所以有时为了确定顶点的入度,可以将图用逆邻接表存储,即对图中每个顶点 i 建立一个单链表,把顶点 i 邻接的顶点存放在一个链表中,即边表中存放的是以 vi 为起始点的边。

有向图的逆邻接表如图 5.2(c)所示。

○邻接表表示法的特点:图的邻接表表示不惟一。

○邻接表表示法的优点:容易找任一结点的第一邻接点和下一个邻接点。存储量小,在存储稀疏图时,比用邻接矩阵节省空间,特别是当和边相关的信息较多的情况下更是如此。

○邻接表表示法的缺点:判定任意两个结点之间是否有边不方便。

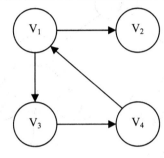

(a)有向图 G

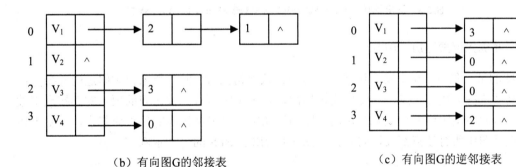

(b)有向图 G 的邻接表　　　　　　　(c)有向图 G 的逆邻接表

图 5.2　图及其邻接表与逆邻接表

5.2.4 图的遍历

5.2.4.1 定义

图的遍历是指从图中的某个顶点(称为初始点)出发,按照某种搜索方法沿着图中的边访问图中的所有顶点,且每个顶点只被访问一次,这个过程称为图的遍历。

图的遍历算法是求解图的连通性问题、拓扑排序和求关键路径等算法的基础。

图的遍历通常有两种方法：深度优先搜索和广度优先搜索（也称宽度优先探索），它们对无向图和有向图都适用。这两种算法的时间复杂度相同，不同之处仅仅在于对顶点访问的顺序不同。

5.2.4.2 深度优先搜索遍历（DFS）

类似于树的先（根）序遍历，它的基本思想是：从图的某一顶点 V0 出发，访问此顶点；然后依次从 V0 的未被访问的邻接点出发，深度优先遍历图，直至图中所有和 V0 邻接的顶点都被访问到；若此时图中尚有顶点未被访问，则另选图中一个未被访问的顶点作起点，重复上述过程，直至图中所有顶点都被访问为止。图 5.3 给出了 DFS 的一个示例。

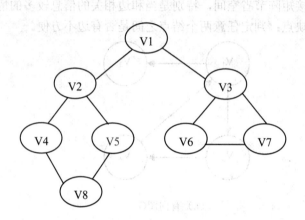

图5.3　深度遍历：V1⇒ V2 ⇒V4 ⇒ V8 ⇒V5 ⇒V3 ⇒V6 ⇒V7

5.2.4.3 广度优先搜索遍历（BFS）

从图的某一顶点 V0 出发，访问此顶点后，依次访问 V0 的各个未曾访问过的邻接点；然后分别从这些邻接点出发，广度优先遍历图，直至图中所有已被访问的顶点的邻接点都被访问到；若此时图中尚有顶点未被访问，则另选图中一个未被访问的顶点作起点，重复上述过程，直至图中所有顶点都被访问为止。图 5.4 给出了 BFS 的一个示例。

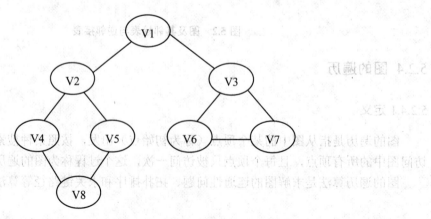

图5.4　广度遍历：V1⇒ V2 ⇒V3 ⇒ V4 ⇒V5 ⇒V6 ⇒V7 ⇒V8

5.2.5 图的基本运算

1) CreateGraph（&G，V，VR）构造：按 V 所给的顶点集和 VR 所给的边集构造一个新图 G。
2) DestroyGraph（&G）销毁：销毁图 G。
3) LocateVex（G，v）顶点定位：若 G 中存在顶点 v，则返回其在图中的位置。
4) GetVex（G，v）取顶点值：返回图 G 中顶点 v 的值。
5) PutVex（&G，v，value）赋值：将值 value 赋给图 G 中顶点 v。
6) FirstAdjVex（G，v）取第一个邻接点：返回图 G 中顶点 v 的第一个邻接点，若其无邻接点则返回 NULL。
7) NextAdjVex（G，v，w）取下一个邻接点：返回图 G 中顶点 v 相对于顶点 w 的下一个邻接点，若 w 是 v 的最后一个邻接点则返回 NULL。
8) InsertVex（&G，v）插入顶点：在图 G 中增加新顶点 v。
9) DeleteVex（&G，v）删除顶点：在图 G 中删除顶点 v 及与其相关的边。
10) InsertArc（&G，v，w）插入边：在图 G 中增加一条从顶点 v 到顶点 w 的边。
11) DeleteArc（&G，v，w）删除边：在图 G 中删除一条从顶点 v 到顶点 w 的边。
12) DFSTraverse（G，v，visit（））深度优先遍历：从顶点 v 出发用 visit（）深度优先遍历图 G。
13) BFSTraverse（G，v，visit（））广度优先遍历：从顶点 v 出发用 visit（）广度优先遍历图 G。

5.3 实践案例

5.3.1 公交路线管理模拟系统

5.3.1.1 系统简介

本项目是对公交车线路信息的简单模拟，以完成建立公交路线信息、修改公交路线信息和删除公交路线信息等功能。

5.3.1.2 设计思路

本项目的实质是完成对公交线路信息的建立、查找、插入、修改、删除等功能，可以首先定义项目的数据结构，然后将每个功能写成一个函数来完成对数据的操作，最后完成主函数以验证各个函数功能并得出运行结果。

○ **数据结构**

公交站点之间的关系可以是任意的，任意两个站点之间都可能相关。而在图形结构中，结点之间的关系可以是任意的，图中任意两个数据元素之间都可能相关。所以可以用图形结构来表示 n 个公交站点之间以及站点之间可能设置的公交路线，其中网的顶点表示公交站点，边表示两个站点之间的路线，赋予边的权值表示相应的距离。

因为公交路线是有一定的连续关系的，如果想输出从某一个起始点开始到某一终点结束的公交路线，就需要找到从某一顶点开始的第一个邻接点和下一个邻接点。因为在邻接表中容易找到任一顶点的第一个邻接点和下一个邻接点，所以本项目使用了图的邻接表存储结构。

○头文件定义

站点类 Node，站点之间的路径类 Linklist 和公交车站模拟类 Station 在头文件 Station.h 中的具体定义为：

```
/***************************************************************
        公交站点类
***************************************************************/
class Node
{
    friend    class    Linklist;
    friend    class    Station;
public:
    Node()
    {
        name = '?';
        leg = 0;
        next = NULL;
    }

private:
    string    name;                              //可到达站点名称
    float     leg;                               //可到达站点路线长度
    Node      *next;                             //站点指针
};


/***************************************************************
        公交站点之间的路线类
***************************************************************/
class Linklist
{
    friend class Station;
public:
        void       Creat     ( Station &L, Linklist *p );        //初始化路线长度
        static void   Check     ( Linklist *p, string name );       //查找路线长度
        static void   Delete    ( Linklist *p, string name );       //删除两站间的路线
        static void   Insert    ( Linklist *p, string name );       //在两站之间插入路线
        static bool   IsT       ( Linklist *p, string name );       //判断两站之间是否存在路线
        static void   Update    ( Linklist *p, string name );       //更改路线长度
    Linklist()
    {
```

```cpp
        s_name = '?';
        sta_name = NULL;
        next_link = NULL;
    }

private:
    string   s_name;                                //本站站点名称
    Node     *sta_name;                             //可到达站点指针
    Linklist *next_link;                            //链表指针
};


/********************************************************************
    公交车站类
********************************************************************/
class Station
{
    friend class Linklist;
public:
    void    Check_path      ( Station &L );         //查找路线长度
    void    Creat_path      ( Station &L );         //初始化路线
    void    Creat_station   ( Station &L );         //初始化站点
    void    Delete_path     ( Station &L );         //删除两站间的路线
    void    Insert_path     ( Station &L );         //在两站之间插入路线
    bool    IsT             ( Station &L, string name ); //判断两站之间是否存在路线
    void    Print           ( Station &L );         //输出所有站点和路线
    int     Stat            ( Station &L );         //统计站点个数
    void    Update_path     ( Station &L );         //更改路线长度

private:
    Linklist *head;                                 //线路图的对外接口指针
};
```

5.3.1.3 程序清单

```cpp
/********************************************************************
    查找路线长度
********************************************************************/
void Linklist::Check(Linklist *p, string name)
{
    if(Linklist::IsT(p, name)) {
        Node *N = p->sta_name;
```

```cpp
        while(N->name != name) {
            N = N->next;
        }
        cout << "由" << p->s_name << "到" << N->name;
        cout << "的路线长度是:  " << N->leg << '\n';
    }
    else {
        cout << "两站点之间不存在路线！ " << '\n';
    }
}

/************************************************************************
    初始化路线长度
************************************************************************/
void Linklist::Creat(Station &L, Linklist *p)
{
    cout << "\n 请输入由" << p->s_name << "站可到达的其他车站个数: ";
    int n;
    cin >> n;
    if (n >= 0 && n < L.Stat(L)) {
        cout << "请依次输入可到达的车站名字和路程: " << '\n';
        for(int i = 1; i <= n; i ++)
        {
            Node *N = new Node;
            cout << "站名: ";
            string num;
            cin >> num;
            if (L.IsT(L, num) && p->s_name != num && !Linklist::IsT(p, num)) {
                N->name = num;
                cout << "路程: ";
                float len;
                cin >> len;
                N->leg = len;
                N->next = p->sta_name;
                p->sta_name = N;
            }
            else {
                cout << "错误的站点, 请重新输入！ " << '\n';
                i --;
            }
        }
```

            }
        }
        else {
            cout << "站数超过车站总数，请重新输入！" << '\n';
            Linklist::Creat(L, p);
        }
}
/**
    删除两站间的路线
**/
void Linklist::Delete(Linklist *p, string name)
{
    Node *N = p->sta_name;
    Node *s = N;
    while (N && N->name != name) {
        s = N;
        N = N->next;
    }
    if (N) {
        if (N == s) {
            p->sta_name = s->next;
        }
        else
            s->next = N->next;
        delete N;
    }
    else {
        cout << "两站之间无路线！" << '\n';
    }
}
/**
    在两站之间插入路线
**/
void Linklist::Insert(Linklist *p, string name)
{
    if(!Linklist::IsT(p, name)) {
        cout << "请输入由" << p->s_name << "到" << name << "的路线长度：";
        float len;
        cin >> len;
        Node *N = new Node;

```cpp
            N->name = name;
            N->leg = len;
            N->next = p->sta_name;
            p->sta_name = N;
        }
        else {
            cout << "两站点之间已存在路线,无法插入新的路线!" << '\n';
        }
    }

/**************************************************************************
    判断两站之间是否存在路线
**************************************************************************/
bool Linklist::IsT(Linklist *p, string name)
{
    Node *N = p->sta_name;
    while(N){
        if(N->name == name)
            return true;
        N = N->next;
    }
    delete N;
    return false;
}


/**************************************************************************
    更改路线长度
**************************************************************************/
void Linklist::Update(Linklist *p, string name)
{
    if(Linklist::IsT(p, name)) {
        cout << "请输入新的由" << p->s_name << "到" << name << "的路线长度:";
        float len;
        cin >> len;
        Node *N = p->sta_name;
        while(N->name != name) {
            N = N->next;
        }
        N->leg = len;
    }
```

```cpp
    else {
        cout << "两站点之间不存在路线，无法修改路线！" << '\n';
    }
}

/*************************************************************************
    初始化站点
*************************************************************************/
void Station::Creat_station(Station &L)
{
    cout << "请输入车站的个数：";
    int n;
    cin >> n;
    cout << "请依次输入车站的名称：" << '\n';
    Linklist *s = new Linklist;
    for(int i = 1; i <= n; i ++)
    {
        cout << "第" << i <<  "个车站名：";
        Linklist *p = new Linklist;
        string name;
        cin >> name;
        p->s_name = name;
        p->next_link = NULL;
        if (i == 1) {
            L.head = p;
            s = L.head;
        }
        else {
            s->next_link = p;
            s = p;
        }
    }
}

/*************************************************************************
    初始化路线
*************************************************************************/
void Station::Creat_path(Station &L)
{
    Linklist *p = L.head;
```

```cpp
        while (p) {
            p->Creat(L, p);
            p = p->next_link;
        }
        delete p;
    }
    /***************************************************************************
        在两站之间插入路线
    ***************************************************************************/
    void Station::Insert_path(Station &L)
    {
        cout << "请输入要添加的路线的两个站点：";
        string name1, name2;
        cin >> name1 >> name2;
        if (Station::IsT(L, name1) && Station::IsT(L, name2)) {
            Linklist *p = L.head;
            while (p) {
                if (p->s_name == name1)
                    Linklist::Insert(p, name2);

                else if (p->s_name == name2)
                    Linklist::Insert(p, name1);

                p = p->next_link;
            }
            delete p;
        }
        else {
            cout << "错误的站点，请重新输入！" << '\n';
            Station::Insert_path(L);
        }
    }


    /***************************************************************************
        查找路线长度
    ***************************************************************************/
    void Station::Check_path(Station &L)
    {
        cout << "请输入要查找的路线的两个站点：";
        string name1, name2;
```

```cpp
        cin >> name1 >> name2;
        if (Station::IsT(L, name1) && Station::IsT(L, name2)) {
            Linklist *p = L.head;
            while (p) {
                if (p->s_name == name1)
                    Linklist::Check(p, name2);

                else if (p->s_name == name2)
                    Linklist::Check(p, name1);

                p = p->next_link;
            }
            delete p;
        }
        else {
            cout << "错误的站点，请重新输入！" << '\n';
            Station::Check_path(L);
        }
    }

    /***************************************************************************
        更改路线长度
    ***************************************************************************/
    void Station::Update_path(Station &L)
    {
        cout << "请输入要修改的路线的两个站点：";
        string name1, name2;
        cin >> name1 >> name2;
        if (Station::IsT(L, name1) && Station::IsT(L, name2)) {
            Linklist *p = L.head;
            while (p) {
                if (p->s_name == name1)
                    Linklist::Update(p, name2);

                else if (p->s_name == name2)
                    Linklist::Update(p, name1);

                p = p->next_link;
            }
            delete p;
```

```cpp
        }
        else {
            cout << "错误的站点，请重新输入！" << '\n';
            Station::Update_path(L);
        }
}


/************************************************************************
    删除两站间的路线
************************************************************************/
void Station::Delete_path(Station &L)
{
    cout << "请输入要删除的路线的两个站点：";
    string name1, name2;
    cin >> name1 >> name2;
    if (Station::IsT(L, name1) && Station::IsT(L, name2)) {
        Linklist *p = L.head;
        while (p) {
            if (p->s_name == name1)
                Linklist::Delete(p, name2);

            else if (p->s_name == name2)
                Linklist::Delete(p, name1);

            p = p->next_link;
        }
        delete p;
    }
    else {
        cout << "错误的站点，请重新输入！" << '\n';
        Station::Delete_path(L);
    }
}

/************************************************************************
    判断两站之间是否存在路线
************************************************************************/
bool Station::IsT(Station &L, string name)
{
    Linklist *p = L.head;
```

```cpp
        while (p) {
            if (p->s_name == name) {
                return true;
            }
            p = p->next_link;
        }
        delete p;
        return false;
    }

    /***************************************************************************
        输出所有站点和路线
    ***************************************************************************/
    void Station::Print(Station &L)
    {
        cout << "所有站点和路线为：" << endl;
        Linklist *p = L.head;
        while (p) {
            Node *N = p->sta_name;
            cout << p->s_name;
            while (N) {
                cout << "-->" << N->name << '(' << N->leg << ')';
                N = N->next;
            }
            cout << '\n';
            delete N;
            p = p->next_link;
        }
        cout << endl;
        delete p;
    }

    /***************************************************************************
        统计站点个数
    ***************************************************************************/
    int Station::Stat(Station &L)
    {
        int i = 0;
        Linklist *p = L.head;
        while (p) {
```

```cpp
            i ++;
            p = p->next_link;
        }
        delete p;
        return i;
    }

/***********************************************************************
    主函数
***********************************************************************/
void main ()
{
    cout<<"\n**              公交车站模拟系统                    **"<<endl;
    cout<<"======================================================"<<endl;
    cout<<"**              A --- 创建公交车站                  **"<<endl;
    cout<<"**              B --- 创建公交路线                  **"<<endl;
    cout<<"**              C --- 检查公交路线                  **"<<endl;
    cout<<"**              D --- 删除公交路线                  **"<<endl;
    cout<<"**              I --- 插入公交路线                  **"<<endl;
    cout<<"**              U --- 更新公交路线                  **"<<endl;
    cout<<"**              E --- 退出    程序                  **"<<endl;
    cout<<"======================================================"<<endl;
    Station L;
    char ch = ' ';
    while( ch!='E' ){
        cout<< '\n' <<"请选择操作 : ";
        cin >> ch;
        switch(ch) {
        case 'A':
            {
                L.Creat_station(L);
                break;
            }
        case 'B':
            {
                L.Creat_path(L);
                L.Print(L);
                break;
            }
        case 'C':
```

```cpp
            {
                L.Check_path(L);
                break;
            }
        case 'D':
            {
                L.Delete_path(L);
                L.Print(L);
                break;
            }
        case 'I':
            {
                L.Insert_path(L);
                L.Print(L);
                break;
            }
        case 'U':
            {
                L.Update_path(L);
                L.Print(L);
                break;
            }
        case 'E':
            break;
        default:
            {
                cout << "输入错误,请选择正确的操作!" << '\n';
                break;
            }
        }
    }
}
```

5.3.1.4 运行结果

```
******          公交车站模拟系统              ******
=================================================
**              A --- 创建公交车站              **
**              B --- 创建公交路线              **
**              C --- 检查公交路线              **
**              D --- 删除公交路线              **
**              I --- 插入公交路线              **
**              U --- 更新公交路线              **
**              E --- 退出  程序                **
=================================================

请选择操作：A
请输入车站的个数：3
请依次输入车站的名称：
第1个车站名：a
第2个车站名：b
第3个车站名：c

请选择操作：B

请输入由a站可到达的其它车站个数：2
请依次输入可到达的车站名和路程：
站名：b
路程：25
站名：c
路程：12

请输入由b站可到达的其它车站个数：2
请依次输入可到达的车站名和路程：
站名：a
路程：30
站名：c
路程：8

请输入由c站可到达的其它车站个数：1
请依次输入可到达的车站名和路程：
站名：a
路程：18
所有站点和路线为：
a-->c(12)-->b(25)
b-->c(8)-->a(30)
c-->a(18)
```

```
请选择操作：C
请输入要查找的路线的两个站点：a c
由a到c的路线长度是：12
由c到a的路线长度是：18

请选择操作：D
请输入要删除的路线的两个站点：a b
所有站点和路线为：
a-->c(12)
b-->c(8)
c-->a(18)

请选择操作：I
请输入要添加的路线的两个站点：a b
请输入由a到b的路线长度：25
请输入由b到a的路线长度：28
所有站点和路线为：
a-->b(25)-->c(12)
b-->a(28)-->c(8)
c-->a(18)

请选择操作：U
请输入要修改的路线的两个站点：a c
请输入新的由a到c的路线长度：15
请输入新的由c到a的路线长度：19
所有站点和路线为：
a-->b(25)-->c(15)
b-->a(28)-->c(8)
c-->a(19)

请选择操作：E
Press any key to continue_
```

5.3.2 最短路径导航查询系统

5.3.2.1 系统简介

设计一个交通导航咨询系统，能让旅客咨询从任一个城市顶点到另一个城市顶点之间的最短路径问题。设计分三个部分，一是建立交通网络图的存储结构；二是解决单源最短路径问题；最后再实现两个城市顶点之间的最短路径问题。

随着计算机的普及以及地理信息科学的发展，GIS（地理信息系统）因其强大的功能得到日益广泛和深入的应用。网络分析作为 GIS 最主要的功能之一，在电子导航、交通旅游、城市规划以及电力、通讯等各种管网、管线的布局设计中发挥了重要的作用。而网络分析中最基本和关键的问题是最短路径问题。

最短路径不仅仅指一般地理意义上的距离最短，还可以引申到其他的度量，如时间、费用、线路容量等。相应地，最短路径问题也可以成为最快路径问题、最低费用问题，等等。由于最短路径问题在实际中常用于汽车导航系统以及各种应急系统（110 报警、119 火警以及医疗救护系统），这些系统一般要求计算出到出事地点的最佳路线的时间一般在 1～3s，在行车过程中还需要实时计算出车辆前方的行驶路线，这就决定了最短路径问题的实现应该是高效率的。最优路径问题不仅包括最短路径问题，还有可能涉及最少时间问题、最少收费（存在收费公路）问题或者是几个问题的综合，这时必须考虑道路级别、道路流量、道路穿越代价（如红灯平均等待时间）等诸多因素。但是必须指出的是，一般来说最优路径在距离上应该是最短的，但最短路径在行驶时间和能源消耗的意义上未必是最优的。其实，无论是距离最短、时间最快还是费用最低，它们的核心算法都是最短路径算法。

5.3.2.2 设计思路

单源最短路径算法的主要代表之一是 Dijkstra（迪杰斯特拉）算法。该算法是目前多数系统解决最短路径问题采用的理论基础，在每一步都选择局部最优解，以期望产生一个全局最优解。

Dijkstra 算法的基本思路是：对于图 $G=(V, E)$，V 是包含 n 个顶点的顶点集，E 是包含 m 条弧的弧集，(v, w) 是 E 中从 v 到 w 的弧，$c(v, w)$ 是弧 (v, w) 的非负权值，设 s 为 V 中的顶点，t 为 V 中可由 s 到达的顶点，则求解从 s 至 t 具有最小弧权值和的最短路径搜索过程如下：

（1）将 v 中的顶点分为 3 类：已标记点、未标记点、已扫描点。将 s 初始化为已标记点，其他顶点为未标记点。为每个顶点 v 都建立一个权值 d 和后向顶点指针 p，并将 d 初始化如下：$d(v)=0, v=s$；$d(v)=\infty, v\neq s$。

（2）重复进行扫描操作：从所有已标记点中选择一个具有最小权值的顶点 v 并将其设为已扫描点，然后检测每个以 v 为顶点的弧 (v, w)，若满足 $d(v)+c(v, w)<d(w)$ 则将顶点 v 设为已标记点，并令 $d(w)=d(v)+c(v, w)$，$p(w)=v$。

（3）若终点 t 被设为已扫描点，则搜索结束。由 t 开始遍历后向顶点指针 P 直至起点 s，即获得最短路径解。

○ **数据结构**

定义一个数组 Map，存储任意两点之间边的权值。

定义一个数组 dist，每个数组元素表示当前所找到的从始点 vi 到每个终点 vj 的最短路径的长度。它的初态为：若从 vi 到 vj 有边，则为边的权值；否则置为 ∞。

定义一个数组 path，其元素 path[k]（$0\leq k\leq n-1$）用以记录 vi 到 vk 最短路径中 vk 的直接前驱结点序号，如果 vi 到 vk 存在边，则 path[k] 初值为 i。

定义一个数组 tag，记录某个顶点是否已计算过最短距离，如果 tag[k]=0，则 vk∈V-S，否则 vk∈S。初始值除 tag[i]=1 以外，所有值均为 0。

○ **程序设计**

查找 min（dist [j], j∈V-S），设 dist[k] 最小，将 k 加入 S 中。

修改对于 V-S 中的任一点 vj，dist[j]=min（dist[k]+w[k][j], dist[j]）且 path[j]=k。

重复直到 V-S 为空。

5.3.2.3 程序清单

```
#include "iostream"
#include "string"
using namespace std;
#define MaxInt 100000
```

/***
    判断图的结点是否存在
***/
static bool IsT(string *name, string nam, int n)
{
    int i = 0;
    while (i < n) {
        if (name[i] == nam) {
            return true;
        }
        i ++;
    }
    return false;
}

/***
    查找结点的存储位置
***/
int Check(string *name, string nam, int n)
{
    int i = 0;
    while (i < n && name[i] != nam) {
        i ++;
    }
    return i;
}

/***
    根据结点和权值构造图
***/
void Creat_path(float *Map, string *name, int n)
{
    cout << "请输入要添加路径的两个结点的名称：";
    string name1, name2;
    cin >> name1 >> name2;
    if (name1 != "?" && name2 != "?") {
        int i, j;
        if (IsT(name, name1, n)) {
            i = Check(name, name1, n);
            j = Check(name, name2, n);

```
            }
            else {
                cout << "错误的输入！" << '\n';
            }
            float weight;
            cout << "请输入权值：";
            cin >> weight;
            Map[n * i + j] = weight;
            Creat_path(Map, name, n);
        }
}


/************************************************************************
    Dijkstra 算法求单源最短路径
*************************************************************************/
float * Dijkstra(float *Map, int *path, int n, int v)
{
    float min;
    int u;
    float *dist = new float[n];
    int *tag = new int [n];
    for(int i = 0; i < n; i ++)
    {
        dist[i] = Map[n * v + i];
        tag[i] = 0;
        if (i != v && dist[i] < MaxInt) {
            path[i] = v;
        }
        else
            path[i] = -1;
    }
    tag[v] = 1;
    for(i = 0; i < n - 1; i ++)
    {
        min = MaxInt;
        u = v;
        for(int j = 0; j < n; j ++)
        {
            if (!tag[j] && dist[j] < min) {
```

```cpp
                    u = j;
                    min = dist[j];
                }
            }
            tag[u] = 1;
            for(int t = 0; t < n; t ++)
            {
                if (!tag[t] && dist[u] + Map[n * u + t] < dist[t]) {
                    dist[t] = dist[u] + Map[n * u + t];
                    path[t] = u;
                }
            }
        }
        return dist;
}

/************************************************************************
    输出最短路径
************************************************************************/
void print(int *path, string *name, int v, int i)
{
    if (path[i] >= 0) {
        print(path, name, v, path[i]);
        cout << name[path[i]] << "-->" ;
    }
}

/************************************************************************
    主函数
************************************************************************/
void main()
{
    cout << "请输入该图的结点数：";
    int n;
    cin >> n;

    float *dist;
    float *Map = new float[n * n];
    for (int i = 0; i < n * n; i ++)
        Map[i] = MaxInt;
```

```
cout << "请依次输入各结点的名称： " << '\n';
string *name = new string[n];
for (i = 0; i < n; i ++)
{
    string nam;
    cin >> nam;
    if (!IsT(name, nam, n)) {
        name[i] = nam;
        Map[n * i + 1] = 0;
    }
    else {
        cout << "错误的输入！ " << '\n';
        i --;
    }
}
cout << endl;


Creat_path(Map, name, n);
int *path = new int [n];

string start;
while(1)
{
    cout << "\n 请输入源点名称：";
    cin >> start;
    if (start=="end")
    {
        break;
    }
    else
    {
        cout << '\n'<<start << "点到图中各点的最短路径： \n" ;
        dist = Dijkstra(Map, path, n, Check(name, start, n));
        for(i = 0; i < n; i ++)
        {
            if(dist[i]==MaxInt || dist[i]==0)
            {
                cout << start << "  点到  " << name[i] << "点无路";
            }
            else
```

```cpp
            {
                cout << start << "点到" << name[i] << "点的最短路径长度为："; 
                cout << dist[i] << "\t\t";
                cout << "最短路径为：";
                print(path, name, Check(name, start, n), i);
                cout << name[i];
            }
            cout << '\n';
        }
        cout << endl;
    }
}
```

5.3.2.4 运行结果

```
请输入该图的结点数: 4
请依次输入各结点的名称:
a b c d

请输入要添加路径的两个结点的名称: a b
请输入权值: 6
请输入要添加路径的两个结点的名称: b c
请输入权值: 4
请输入要添加路径的两个结点的名称: c d
请输入权值: 5
请输入要添加路径的两个结点的名称: b d
请输入权值: 8
请输入要添加路径的两个结点的名称: a d
请输入权值: 18
请输入要添加路径的两个结点的名称: ? ?

请输入源点名称: a

a 点到图中各点的最短路径:
a 点到 a 点无路
a 点到 b 点的最短路径长度为: 6         最短路径为: a-->b
a 点到 c 点的最短路径长度为: 10        最短路径为: a-->b-->c
a 点到 d 点的最短路径长度为: 14        最短路径为: a-->b-->d

请输入源点名称: b

b 点到图中各点的最短路径:
b 点到 a 点无路
b 点到 b 点无路
b 点到 c 点的最短路径长度为: 4         最短路径为: b-->c
b 点到 d 点的最短路径长度为: 8         最短路径为: b-->d

请输入源点名称: c

c 点到图中各点的最短路径:
c 点到 a 点无路
c 点到 b 点无路
c 点到 c 点无路
c 点到 d 点的最短路径长度为: 5         最短路径为: c-->d

请输入源点名称: end
Press any key to continue
```

5.3.3 电网建设造价模拟系统

5.3.3.1 系统简介

假设一个城市有 n 个小区，要实现 n 个小区之间的电网都能够相互接通，构造这个城市 n 个小区之间的电网，使总工程造价最低。请设计一个能满足要求的造价方案。

5.3.3.2 设计思路

在每个小区之间都可以设置一条电网线路，都要付出相应的经济代价。n 个小区之间最多可以有 n（n-1）/2 条线路，选择其中的 n-1 条使总的耗费最少。

○**数据结构**

可以用连通网来表示 n 个城市之间以及 n 个城市之间可能设置的电网线路，其中网的顶点表示小区，边表示两个小区之间的线路，赋予边的权值表示相应的代价。对于 n 个顶点的连通网可以建立许多不同的生成树，每一颗生成树都可以是一个电路网。现在，我们要选择总耗费最少的生成树，就是构造连通网的最小代价生成树的问题，一棵生成树的代价就是树上各边的代价之和。

可以用带权的无向图（即无向网）表示这 n 个小区之间的电网连接，其中顶点表示小区，权值表示城市之间电网建设的造价，构造一个无向网的最小生成树即是满足要求的最低电网连接造价方案。

设 G=（V，E）是具有 n 个顶点的网络，T=（U，TE）为 G 的最小生成树，U 是 T 的顶点集合，TE 是 T 的边集合。

○**程序设计**

Prim 算法的基本思想是：首先从集合 V 中任取一顶点（例如去顶点 v0）放入集合 U 中，这时 U={ v0}，TE=NULL。然后找出所有一个顶点在集合 U 里，另一个顶点在集合 V-U 里的边，使权（u，v）（u∈U，v∈V-U）最小，将该边放入 TE，并将顶点 v 加入集合 U。重复上诉操作直到 U=V 为止。这时 TE 中有 n-1 条边，T=（U，TE）就是 G 的一颗最小生成树。

在构造最小生成树的过程中定义一个类型为 Edge 的数组 mst: Edge mst[n-1]；其中，n 为网络中顶点的个数，算法结束时，mst 中存放求出的最小生成树的 n-1 条边。

○**头文件定义**

电网顶点类 Node，电网顶点间的线路类 Linklist 和电网类 Circuit 在头文件 Circuit.h 中的具体定义为：

```
/******************************************************************
    电网顶点类
******************************************************************/
class Node
{
    friend    class    Linklist;
    friend    class    Circuit;
public:
```

```cpp
        Node();

private:
    bool        flag;                           //访问标志
    string      s_name;                         //出发顶点名称
    string      name;                           //可到达接点名称
    float       leg;                            //可到达顶点边长度
    Node        *next;                          //顶点指针
};
```

/**
    电网顶点间的线路类
**/
```cpp
class Linklist
{
    friend    class    Circuit;
public:
        void         Print      ( Node *N );                              //输出普利姆最小生成树
        static void  Insert     ( Linklist *p, string name, float path ); //在两顶点之间插入边
        static bool  IsT        ( Linklist *p, string name );             //判断两顶点之间是否存在边
        static int   Stat       ( Linklist &L );                          //统计某顶点可到达其他顶点的个数

        Linklist()
        {
            s_name = '?';
            sta_name = NULL;
            next_link = NULL;
        }

        string      s_name;                     //本顶点名称
        Node        *sta_name;                  //可到达顶点指针
        Linklist    *next_link;                 //链表指针
};
```

/**
    电网类
**/
```cpp
class Circuit
{
    friend class Linklist;
```

```cpp
public:
    void        Change_flag   ( Circuit &L, string name );         //更改访问标志
    Node      * Min_path      ( Circuit L, Linklist &edge, string name );  //查找当前顶点的最短边
    void        Insert_path   ( Circuit &L );                      //在顶点之间插入边
    void        Creat_Circuit ( Circuit &L );                      //初始化顶点
    bool        IsT           ( Circuit &L, string name );         //判断两顶点之间是否存在边
    void        Min_Tree      ( Circuit L, Linklist &edge, string name );  //计算最小生成树
    int         Stat          ( Circuit &L );                      //统计顶点个数

private:
    Linklist  *head;                                                //电网的对外接口指针
};
```

5.3.3.3 程序清单

```cpp
#define MAX 65535

/************************************************************************
    顶点类的构造函数
************************************************************************/
Node::Node()
{
    flag = 0;
    name = '?';
    leg = 0;
    next = NULL;
}

/************************************************************************
    在两顶点之间插入边
************************************************************************/
void Linklist::Insert(Linklist *p, string name, float path)
{
    if(!Linklist::IsT(p, name)) {
        Node *N = new Node;
        N->name = name;
        N->leg = path;
        N->next = p->sta_name;
        p->sta_name = N;
    }
    else {
```

```cpp
        cout << "两顶点之间已存在边，无法插入新的边！" << '\n';
    }
}


/************************************************************************
    判断顶点之间是否存在边
************************************************************************/
bool Linklist::IsT(Linklist *p, string name)
{
    Node *N = p->sta_name;
    while(N){
        if(N->name == name)
            return true;
        N = N->next;
    }
    delete N;
    return false;
}


/************************************************************************
    输出普利姆最小生成树
************************************************************************/
void Linklist::Print(Node *N)
{
    if (N->next) {
        Linklist::Print(N->next);
        cout << N->s_name << "-("<< N->leg <<")->"<< N->name << '\t';
    }
}


/************************************************************************
    统计某顶点可到达其他顶点的个数
************************************************************************/
int Linklist::Stat(Linklist &L)
{
    int i = 0;
    Node *N = L.sta_name;
    while (N) {
        N = N->next;
```

```cpp
            i ++;
        }
        return i;
    }
/**************************************************************************
    初始化顶点
**************************************************************************/
void Circuit::Creat_Circuit(Circuit &L)
{
    cout << "请输入顶点的个数：";
    int n;
    cin >> n;
    cout << "请依次输入各顶点的名称："<< '\n';
    Linklist *s = new Linklist;
    for(int i = 1; i <= n; i ++)
    {
        Linklist *p = new Linklist;
        string name;
        cin >> name;
        p->s_name = name;
        p->next_link = NULL;
        if (i == 1) {
            L.head = p;
            s = L.head;
        }
        else {
            s->next_link = p;
            s = p;
        }
    }
}

/**************************************************************************
    在两顶点之间插入边
**************************************************************************/
void Circuit::Insert_path(Circuit &L)
{
    cout << "请输入两个顶点及边：";
    string name1, name2;
    float path;
```

138

```cpp
        cin >> name1 >> name2 >> path;
        if (name1 != "?" && name2 != "?" && path!=0)
        {
            if (Circuit::IsT(L, name1) && Circuit::IsT(L, name2))
            {
                Linklist *p = L.head;
                while (p)
                {
                    if (p->s_name == name1)
                        Linklist::Insert(p, name2, path);

                    else if (p->s_name == name2)
                        Linklist::Insert(p, name1, path);

                    p = p->next_link;
                }
                delete p;
            }
            else {
                cout << "错误的顶点,请重新输入! " << '\n';
            }
            Circuit::Insert_path(L);
        }
}
/******************************************************************************
    判断是否存在此顶点
******************************************************************************/
bool Circuit::IsT(Circuit &L, string name)
{
    Linklist *p = L.head;
    while (p) {
        if (p->s_name == name) {
            return true;
        }
        p = p->next_link;
    }
    delete p;
    return false;
}
```

/***
        更改访问标志
***/
void Circuit::Change_flag(Circuit &L, string name)
{
    Linklist *p = L.head;
    while (p) {
        Node *N = p->sta_name;
        while (N) {
            if (N->name == name) {
                N->flag = 1;
                N->leg = MAX;
                break;
            }
            N = N->next;
        }
        p = p->next_link;
    }
}

/***
        查找当前顶点的最短边
***/
Node * Circuit::Min_path(Circuit L, Linklist &edge, string name)
{
    Linklist *p = L.head;
    while (p->s_name != name) {
        p = p->next_link;
    }
    Node *N = p->sta_name;
    Node *s = N;
    while (N->flag && N->next) {
        N = N->next;
    }
    if (N) {
        s = N;
        while (N) {
            if (s->leg > N->leg && !N->flag) {
                s = N;
            }

```
            N = N->next;
        }
    }
    return s;
}


/************************************************************************
        利用普利姆算法计算最小生成树
*************************************************************************/
void Circuit::Min_Tree(Circuit L, Linklist &edge, string name)
{
    if (Circuit::IsT(L, name)) {
        Linklist *p = L.head;
        while (p->s_name != name) {
            p = p->next_link;
        }

        Node *start = new Node;
        start->name = name;
        start->s_name = name;
        start->leg = 0;
        start->next = edge.sta_name;
        edge.sta_name = start;
        L.Change_flag(L, name);

        while (edge.Stat(edge) < L.Stat(L)) {
            Node *t = edge.sta_name;
            Node *s = t;
            while (t) {
                if ((Circuit::Min_path(L, edge, s->name)->leg > \
                    Circuit::Min_path(L, edge, t->name)->leg) && \
                    !Circuit::Min_path(L, edge, t->name)->flag) {
                    s = t;
                }
                t = t->next;
            }
            Node *N = new Node;
            N->s_name = s->name;
            N->name = Circuit::Min_path(L, edge, s->name)->name;
            N->leg = Circuit::Min_path(L, edge, s->name)->leg;
```

```cpp
                N->next = edge.sta_name;
                edge.sta_name = N;
                L.Change_flag(L, N->name);
            }
        }
        else {
            cout << "错误的输入,请重新输入!"<< '\n';
            Circuit::Min_Tree(L, edge, name);
        }
    }
}

/*************************************************************************
    统计顶点个数
*************************************************************************/
int Circuit::Stat(Circuit &L)
{
    int i = 0;
    Linklist *p = L.head;
    while (p) {
        i ++;
        p = p->next_link;
    }
    delete p;
    return i;
}

/*************************************************************************
    主函数
*************************************************************************/
void main ()
{
    cout<<"\n**              电网造价模拟系统                **"<<endl;
    cout<<"===================================================="<<endl;
    cout<<"**              A --- 创建电网顶点               **"<<endl;
    cout<<"**              B --- 添加电网的边               **"<<endl;
    cout<<"**              C --- 构造最小生成树             **"<<endl;
    cout<<"**              D --- 显示最小生成树             **"<<endl;
    cout<<"**              E --- 退出   程序                **"<<endl;
    cout<<"===================================================="<<endl;
    Circuit L;
```

```cpp
Linklist edge;
char ch = ' ';
while( ch!= 'E' ){
    cout<<'\n'<<"请选择操作 : ";
    cin >> ch;
    switch(ch) {
    case 'A':
        {
            L.Creat_Circuit(L);
            break;
        }
    case 'B':
        {
            L.Insert_path(L);
            break;
        }
    case 'C':
        {
            cout << "请输入起始顶点：";
            string name;
            cin >> name;
            L.Min_Tree(L, edge, name);
            cout << "生成 Prim 最小生成树！\n";
            break;
        }
    case 'D':
        {
            cout << "最小生成树的顶点及边为：" << '\n'<<'\n';
            edge.Print(edge.sta_name);
            cout <<'\n';
            break;
        }
    case'E':
        break;
    default:
        {
            cout << "输入错误，请选择正确的操作！" <<'\n';
            break;
        }
    }

}
```

}

5.3.3.4 运行结果

```
**        电网造价模拟系统        **
================================================
**        A ---- 创建电网顶点        **
**        B ---- 添加电网的边         **
**        C ---- 构造最小生成树       **
**        D ---- 显示最小生成树       **
**        E ---- 退出   程序         **
================================================

请选择操作：A
请输入顶点的个数：4
请依次输入各顶点的名称：
a b c d

请选择操作：B
请输入两个顶点及边：a b 8
请输入两个顶点及边：b c 7
请输入两个顶点及边：c d 5
请输入两个顶点及边：d a 11
请输入两个顶点及边：a c 18
请输入两个顶点及边：b d 12
请输入两个顶点及边：? ? 0

请选择操作：C
请输入起始顶点：a
生成Prim最小生成树！

请选择操作：D
最小生成树的顶点及边为：

a-<8>->b        b-<7>->c        c-<5>->d

请选择操作：E
Press any key to continue_
```

5.3.4 软件工程进度规划系统

5.3.4.1 系统简介

设计一个软件，需要进行用户需求分析、系统需求确认、系统概要设计、设计用例场景、系统的详细设计、数据库详细设计、编码、单元测试、集成测试、系统测试以及系统维护等活动。要求用户需求分析在系统需求确认之前完成，系统的详细设计必须在系统的概要设计、设计系统用例和设计用例场景之前完成。

请判断该软件设计流程是否有回路，并给出该软件设计 AOV 网的拓扑序列。活动之间的具体关系如下表所示。

表 5.1 系统活动之间的关系

活动代码	活动名称	先需活动
A1	用户需求分析	无
A2	系统需求确认	A1
A3	系统概要设计	A2
A4	设计用例场景	无

续表

活动代码	活动名称	先需活动
A5	系统的详细设计	A3，A4
A6	数据库详细设计	A3
A7	编码	A5，A6
A8	单元测试	A7
A9	集成测试	A8
A10	系统测试	A7
A11	维护	A11

5.3.4.2 设计思路

拓扑排序（Topological Sort）是求解网络问题所需的主要算法。管理技术中的计划评审技术 PERT（Performance Evaluation And Review Technique）和关键路径法 CPM（Critical Path Method）都应用这一算法。通常，软件开发、施工过程、生产流程、程序流程等都可作为一个工程。一个工程可分成若干子工程，子工程常称为活动（activity）。因此要完成整个工程，必须完成所有的活动。活动的执行常常伴随着某些先决条件，一些活动必须先于另一些活动被完成。

AOV 网络代表的领先关系应当是一种拟序关系，它具有传递性（transitive）和反自反性（irreflexive）。如果这种关系不是反自反的，就意味着要求一个活动必须在它自己开始之前就完成。这显然是荒谬的，这类工程是不可实施的。如果给定了一个 AOV 网络，我们所关心的事情之一，是要确定由此网络的各边所规定的领先关系是否是反自反的，也就是说，该 AOV 网络中是否包含任何有向回路。一般地，它应当是一个有向无环图（DAG）。

一个拓扑序列（Topological Order）是 AOV 网络中顶点的线性序列，使得对图中任意两个顶点 i 和 j，i 是 j 的前驱结点，则在线性序列中 i 先于 j。

根据需求，设计一个软件的 AOV 网示意图如下图 5.5 所示。

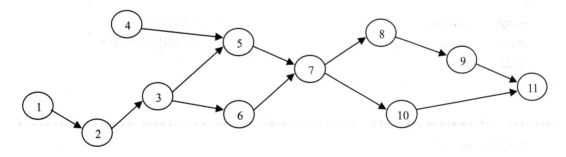

图 5.5 软件设计流程的 AOV 网

○ **数据结构**

拓扑排序可以在不同的存储结构上实现，与遍历运算相似，邻接表方法在这里更有效。拓扑排序算法包括两个基本操作：①决定一个顶点是否入度为零；②删除一个顶点的所有出边。如果我们对每个顶点的直接前驱予以计数，使用一个数组 InDgree 保存每个顶点的入度，即 InDgree[i] 为顶点 i 的入度，则基本操作①很容易实现。而基本操作②在使用邻接表表示时，一般会比邻接矩阵更有效。在邻接矩阵的情况下，必须处理与该顶点有关的整行元素（n 个），而邻接表只需处理在邻接矩阵中非零的那些顶点。

○ 程序设计

拓扑排序算法可描述如下：

（1）在图中选择一个入度为零的顶点，并输出之；

（2）从图中删除该顶点及其所有出边（以该顶点为尾的有向边）；

（3）重复（1）和（2），直到所有顶点都已列出，或者直到剩下的图中再也没有入度为零的顶点为止。后者表示图中包含有向回路。

○ 头文件定义

站点类，站点之间的路径类 Linklist 和 类 在头文件 Station.h 中的具体定义为：

```
/******************************************************************
    拓扑活动类
*******************************************************************/
class Node
{
    friend    class    Linklist;
    friend    class    Topo;
    friend    void     Top_sort ( Topo &T, Linklist &L );
public:
    Node()
    {
        name = '?';
        next = NULL;
    }

private:
    string    s_name;                      //活动名称
    string     name;                       //可到达活动名称
    Node      *next;
};

/******************************************************************
    活动间的序列关系类
*******************************************************************/
class Linklist
{
    friend    class    Topo;
    friend    void     Top_sort ( Topo &T, Linklist &L );
public:
```

```
        void       Print        ( Node *N );                          //输出拓扑排序结果
    static void    Insert       ( Linklist *p, string name );         //在两活动之间插入路径
    static bool    IsT          ( Linklist *p, string name );         //判断两活动之间是否存在路径

    Linklist()
    {
        s_name = '? ';
        start_name = NULL;
        next_link = NULL;
    }
    string     s_name;                                    //本活动名称
    Node       *start_name;                               //可到达活动指针
    Linklist   *next_link;
};

/***************************************************************************
    拓扑序列类
***************************************************************************/
class Topo
{
    friend    class    Linklist;
    friend    void     Top_sort  ( Topo &T, Linklist &L );
public:
        void        Creat_Topo   ( Topo &L );                //初始化活动
        bool        IsT          ( Topo &L, string name );   //判断是否存在此活动
        void        Insert_path  ( Topo &L );                //在活动之间插入路径
    static string   IsHead       ( Topo &L );
    static void     Delete       ( Topo &Temp, string name );

private:
    Linklist *head;
};
```

5.3.4.3 程序清单

```
/***************************************************************************
    在两活动之间插入序列关系
***************************************************************************/
void Linklist::Insert(Linklist *p, string name)
{
    if(!Linklist::IsT(p, name)) {
```

```cpp
        Node *N = new Node;
        N->name = name;
        N->next = p->start_name;
        p->start_name = N;
    }
    else {
        cout << "两活动之间已存在路径，无法插入新的路径！" <<'\n';
    }
}

/************************************************************************
    判断活动之间是否存在序列关系
************************************************************************/
bool Linklist::IsT(Linklist *p, string name)
{
    Node *N = p->start_name;
    while(N){
        if(N->name == name)
            return true;
        N = N->next;
    }
    delete N;
    return false;
}

/************************************************************************
    输出拓扑序列
************************************************************************/
void Linklist::Print(Node *N)
{
    if (N->next) {
        Linklist::Print(N->next);
        cout << "->";
    }
    cout << N->name ;
}

/************************************************************************
    初始化拓扑活动
************************************************************************/
```

```cpp
void Topo::Creat_Topo(Topo &L)
{
    cout << "请输入活动的个数：";
    int n;
    cin >> n;
    cout << "请依次输入活动的名称：" <<'\n';
    Linklist *s = new Linklist;
    for(int i = 1; i <= n; i ++)
    {
        Linklist *p = new Linklist;
        string name;
        cin >> name;
        p->s_name = name;
        p->next_link = NULL;
        if (i == 1) {
            L.head = p;
            s = L.head;
        }
        else {
            s->next_link = p;
            s = p;
        }
    }
}

/************************************************************************
    在两活动之间插入序列关系
************************************************************************/
void Topo::Insert_path(Topo &L)
{
    cout << "为两个活动添加序列：";
    string name1, name2;
    cin >> name1 >> name2;
    if (name1 != "?" && name2 != "?") {
        if (Topo::IsT(L, name1) && Topo::IsT(L, name2)) {
            Linklist *p = L.head;
            while (p) {
                if (p->s_name == name1) {
                    Linklist::Insert(p, name2);
                }
```

```cpp
                p = p->next_link;
            }
            delete p;
        }
        else {
            cout << "错误的活动,请重新输入！" <<'\n';
        }
        Topo::Insert_path(L);
    }
}

/************************************************************************
    判断是否存在此活动
************************************************************************/
bool Topo::IsT(Topo &L, string name)
{
    Linklist *p = L.head;
    while (p) {
        if (p->s_name == name) {
            return true;
        }
        p = p->next_link;
    }
    delete p;
    return false;
}

/************************************************************************
    在有向图中删除活动
************************************************************************/
void Topo::Delete(Topo &Temp, string name)
{
    Linklist *N = Temp.head;
    if (Temp.head->s_name == name) {
        Temp.head = N->next_link;
        delete N;
    }
    else {
        Linklist *s = N;
        while (N) {
```

```cpp
            if (N->s_name == name) {
                s->next_link = N->next_link;
                delete N;
                break;
            }
            s = N;
            N = N->next_link;
        }
    }
}

/************************************************************************
    进行拓扑排序
************************************************************************/
void Top_sort(Topo &T, Linklist &L)
{
    Topo Temp = T;
    string name;
    while (Temp.head) {
        if (Topo::IsHead(Temp) != "false") {
            name = Topo::IsHead(Temp);
        }
        else{
            cout << "该图中存在有向环！"<<'\n';
            return;
        }
        L.Insert(&L, name);
        Topo::Delete(Temp, name);
    }
    L.Print(L.start_name);
    cout << endl;
}

/************************************************************************
    判断该图中是否存在入度为零的活动，并返回
************************************************************************/
string Topo::IsHead(Topo &L)
{
    Linklist *p = L.head;
    while (p) {
        int i = 0;
```

```cpp
            Linklist *temp = L.head;
            while (temp) {
                if (Linklist::IsT(temp, p->s_name)) {
                    i = 1;
                    break;
                }
                temp = temp->next_link;
            }
            if (!i) {
                return p->s_name;
                break;
            }
            p = p->next_link;
        }
        return "false";
    }

/************************************************************************
                            主函数
************************************************************************/
    void main ()
    {
        cout<<"\n**            进度规划模拟系统              **"<<endl;
        cout<<"================================================="<<endl;
        cout<<"**            A --- 创建活动序列              **"<<endl;
        cout<<"**            B --- 添加活动序列              **"<<endl;
        cout<<"**            C --- 拓扑   排序              **"<<endl;
        cout<<"**            E --- 退出   程序              **"<<endl;
        cout<<"================================================="<<endl;
        Topo T;
        Linklist L;
        char ch = '';
        while( ch!= 'E' ){
            cout<<'\n'<<"请选择操作 : ";
            cin >> ch;
            switch(ch) {
            case'A':
                {
                    T.Creat_Topo(T);
                    break;
                }
```

```cpp
        case'B':
            {
                T.Insert_path(T);
                break;
            }
        case'C':
            {
                Top_sort(T, L);
                break;
            }
        case'E':
            break;
        default:
            {
                cout << "输入错误，请选择正确的操作！"<<'\n';
                break;
            }
        }
    }
}
```

5.3.4.4 运行结果

```
**           进度规划模拟系统            **
========================================
**        A ---- 创建活动序列           **
**        B ---- 添加活动序列           **
**        C ---- 拓扑    排序           **
**        E ---- 退出    程序           **
========================================

请选择操作 : A
请输入活动的个数: 11
请依次输入活动的名称:
A1 A2 A3 A4 A5 A6 A7 A8 A9 A10 A11

请选择操作 : B
为两个活动添加序列: A1 A2
为两个活动添加序列: A2 A3
为两个活动添加序列: A3 A5
为两个活动添加序列: A4 A5
为两个活动添加序列: A3 A6
为两个活动添加序列: A5 A7
为两个活动添加序列: A6 A7
为两个活动添加序列: A7 A8
为两个活动添加序列: A7 A10
为两个活动添加序列: A8 A9
为两个活动添加序列: A9 A11
为两个活动添加序列: A10 A11
为两个活动添加序列: ? ?

请选择操作 : C
A1->A2->A3->A4->A5->A6->A7->A8->A9->A10->A11

请选择操作 : E
Press any key to continue
```

5.4 巩固提高

设计一个景区旅游导航系统，实现旅游景点导游线路的最短路径查询。
○要求
（1）创建景区景点并绘制景点分布图；
（2）输出导游线路图；
（3）查询任意两个景点间的最短路径。
○提示
需要计算出所有景点之间最短路径和最短距离，可分别采用迪杰斯特拉算法或弗洛伊德算法。

第六章 查找

在日常生活中，人们几乎每天都要进行查找工作。例如，在通信簿中查阅某单位或某人的通信地址；在字典中查阅某个字的读音和含义等。其中"通信簿"和"字典"都可看作一张查找表。而查找表是一种在实际应用中大量使用的数据结构。查找表是由同一类型的数据元素（或记录）构成的集合。由于"集合"中的数据元素之间存在着完全松散的关系，因此查找表是一种非常灵便的数据结构。

在各种系统软件或应用软件中，查找表也是一种最常见的结构之一。如编译程序中的符号表、信息处理系统中的信息表均是查找表。同样，在计算机中进行查找的方法也随数据结构的不同而不同。由于查找表中数据元素之间仅存在"同属一个集合"的松散关系，因此需要在数据元素之间人为地加上一些关系，以便按某种规则进行查找，即以另一种数据结构来表示查找表。本章分别就线性表、树表和哈希表三种数据结构的查找来讨论查找表的表示和操作实现的方法。

6.1 实践目的和要求

6.1.1 实践目的

1）掌握各种查找算法的思想及其适用条件；
2）掌握上机实现各种查找算法的基本方法；
3）熟练掌握二叉排序树的构造和查找方法；
4）掌握散列存储结构的思想，能选择合适散列函数，实现不同冲突处理方法的散列表的查找、建立；
5）掌握运用不同类型数据结构的查找算法解决应用问题。

6.1.2 实践要求

1）掌握本章实践的算法。
2）上机运行本章的程序，保存和打印出程序的运行结果，并结合程序进行分析。
3）注意理解折半查找的适用条件（如链表能否实现折半查找？）。
4）注意建立二叉排序树、散列表时相同元素的处理。
5）注意理解静态查找、动态查找概念。
6）比较各种查找算法的各自特点，能够根据实际情况选择合适的查找方法。

6.2 基本概念

6.2.1 查找的概念

"查找"也叫检索，是根据给定的某个数据值 k，在含有 n 个记录的表（称为查找表）中找出一个关键字等于 k 的记录。（关键字是记录中某个数据项的值，它可以惟一标识一个记录）。若找到，则查找成功，返回该记录的信息或该记录在查找表中的位置；否则查找失败，返回相关提示信息。

查找方法有多种。

若整个查找过程都在内存进行则称为内查找；若查找过程中需要访问外存，则称为外查找。

若查找目的只是为了确定指定条件的结点存在与否，称为静态查找；若查找是为确定结点的插入位置或为了删除找到的结点，称为动态查找。

在查找时具体采用何种查找方法取决于如下因素：
○ 使用哪种数据结构来表示查找表，即查找表中的记录是按何种方式组织的。
○ 查找表中关键字的次序，即查找表是无序表还是有序表。

6.2.2 线性表的查找

查找与数据的存储结构有关，线性表有顺序和链式两种存储结构，这里只介绍以顺序表作为存储结构时实现的查找算法。这是最常见的查找方式。结点只含关键码，并且是整数。如果线性表无序，则采用顺序查找；如果线性表有序，则可以使用顺序查找或折半查找。

6.2.2.1 顺序查找

顺序查找是一种最基本和最原始的查找方法。它的思路是，从表中的第一个元素开始，将给定的值与表中逐个元素的关键字进行比较，直到两者相符，查到所要找的元素为止。否则就是表中没有要找的元素，查找不成功。对于表中记录的关键字是无序的表，只能采用这种方法。

设表的长度为 n，顺序查找的时间复杂度是 $O(n)$。

6.2.2.2 折半查找（二分法查找）

折半查找又称二分查找，是针对有序表进行查找的简单、高效而又较常用的方法。有序表是指表中的各记录按关键字的值有序（升序或降序，在此假设是升序）存放。

折半查找不像顺序查找那样，从第一个记录开始逐个顺序搜索，其基本思想是：首先选取表中间位置的记录，将其关键字与给定关键字 k 进行比较，若相等，则查找成功；否则，若 k 值比该关键字值大，则要找的元素一定在表的后半部分（或称右子表），继续对右子表进行折半查找；若 k 值比该关键字值小，则要找的元素一定在表的前半部分（左子表），继续对左子表进行折半查找。每进行一次比较，要么找到要查找的元素，要么将查找的范围缩小一半。如此递推，直到查找成功或把要查找的范围缩小为空（查找失败）。

设表的长度为 n，折半查找的平均查找长度（ASL）是 $O(\log_2 n)$，所以其时间复杂度为 $O(\log_2 n)$。

从折半查找的平均查找长度来看，当表的长度 n 很大时，该方法尤其能显示出其时间效率。但由于折半查找的表仍是线性表，若经常要进行插入、删除操作，则元素排序费时太多，因此折半查找比较适合于一经建立就很少改动而又需要经常查找的线性表，较少查找而又经常需要改动的线性表可以采用线性链表存储，使用顺序查找。

6.2.2.3 分块查找

在处理线性表时，如果既希望能够快速查找，又要经常动态变化，则可以采用分块查找方法。分块查找又称索引顺序查找，要求将待查的记录均匀地分成块，块间按大小排序，块内不排序。因此需要建立一个块的最大（或最小）关键字表，称之为"索引表"。

具体而言，假设我们按结点元素关键字升序方式组织表中各块，则要求第一块中任一结点的关键字值都小于第二块中所有结点的关键字值；第二块中任一结点的关键字值都小于第三块中所有结点的关键字值；依此类推。然后选择每块中的最大（或最小）关键字值组成索引表。换言之，索引表中的结点个数等于线性表被分割的块数。

分块查找过程分两步进行，先用二分法在索引表中查索引项，确定要查的结点在哪一块。然后，再在相应块内顺序查找。例如要找关键字为 k 的元素，则先用折半查找法由索引表查出 k 所在的块，再用顺序查找法在对应的块中找出 k。

分块查找的平均查找长度位于顺序查找和折半查找之间。

6.2.2.4 顺序表的三种查找方法的比较

1）平均查找长度：折半查找最小，分块查找次之，顺序查找最大；
2）表的结构：顺序查找对有序表、无序表均适用；折半查找仅适用于有序表；分块查找要求表中元素是逐段有序的，就是块与块之间的记录按关键字有序；
3）存储结构：顺序查找和分块查找均适用于顺序表和线性链表两种存储结构；折半查找只适用于顺序表存储结构。
4）分块查找综合了顺序查找和折半查找的优点，既能较快地查找，又能适应动态变化的要求。

6.2.3 树表的查找

以二叉树或树作为表的组织形式，称为树表。树表在进行插入或删除操作时，可以方便地维护表的有序性，不需要移动表中记录，从而减少因移动记录引起的额外时间开销。树表适于进行动态查找。树表有不同的表示方法。

6.2.3.1 二叉排序树（BST）

1. 二叉排序树（二叉查找树）的定义

二叉排序树或者是一棵空树，或者是具有下列性质的一棵二叉树：
（1）若左子树不空，则左子树上的所有记录的值都小于根记录的值；
（2）若右子树不空，则右子树上的所有记录的值都大于根记录的值；
（3）左右子树本身也分别是二叉排序树。

从 BST 性质可以推出 BST 的另一个性质：按中序遍历该树会得到一个升序序列。

2. 二叉排序树的查找

从 BST 定义可知，在二叉树上进行查找与折半查找类似。查找过程为：若给定关键字值等于根结点的关键字值，则查找成功；若小于根结点的关键字值，则查找根结点的左子树；若大于根结点的关键字值，则查找根结点的右子树。在左右子树上的操作类似。这显然是一个递归过程。

3. 二叉排序树的插入与构造操作

○插入结点的过程：若原二叉排序树为空，则插入结点成为新的根结点；否则，若插入结点的关键字值小于根结点的关键字值，则插入到根结点的左子树中；若插入结点的关键字值大于根结点的关键字值，则插入到根结点的右子树中。

注意：新插入的结点均成为叶子结点。

○二叉排序树的构造：是通过依次输入数据元素，并把它们插入到二叉树的适当位置上。具体过程是：每读入一个元素，建立一个新结点，若二叉排序树为空，则新结点作为二叉排序树的根；否则，将新结点插入到二叉排序树中。

4. 二叉排序树的删除操作

在二叉排序树中删除一个结点，相当于删除有序序列中的一个元素，不能把以该结点为根的子树都删除，只能删除这个根结点并仍保持二叉排序树的特性。

设要删除二叉排序树中的 p 结点，p 的双亲结点为 f，分四种情况：

（1）p 为叶子结点，只需修改 f 的指针：f->lchild=NULL 或 f->rchild=NULL。

（2）p 只有左子树，用 p 的左孩子代替 p。

（3）p 只有右子树，用 p 的右孩子代替 p。

（4）p 的左、右子树均非空，方法一：用 p 的左子树的最右下结点 r 代替 p；方法二：用 p 的右子树的最左下结点 r 代替 p。

就算法的时间复杂度而言，二叉排序树的查找和折半查找差不多。但就插入和删除操作而言，前者更高效，只需修改指针而无须移动记录。二叉排序树的插入和删除操作的时间复杂度均为 $O(\log_2 n)$。

6.2.3.2 平衡二叉树（AVL 树）

平衡二叉树或者是一棵空树，或者是具有下列性质的二叉排序树：其左子树和右子树都是平衡二叉树且左子树和右子树的深度之差不会超过 1。

○注意：完全二叉树必为平衡树，平衡树不一定是完全二叉树。

二叉树上各结点的平衡因子是指结点左子树与其右子树深度之差。可见，平衡二叉树上所有结点的平衡因子只可能是-1、0 和 1。

假设向平衡二叉树中插入一个新结点后破坏了平衡二叉排序树的平衡性，首先要找出插入新结点后失去平衡的最小子树（失去平衡的最小子树指以离插入结点最近，且平衡因子绝对值大于 1 的结点为根的子树）根结点的指针，然后再调整这个子树中有关结点之间的链接关系，使之成为新的平衡子树。当失去平衡的最小子树被调整为平衡子树后，原有其他所有不平衡子树无需调整，整个二叉排序树又成为一棵平衡二叉树。

在插入一个结点后，如果原二叉树失去了平衡，设失去平衡的最小子树的根结点为 A，则可分下列四种情况调整该子树：

（1）LL 型：新结点插入到 A 的左子树中而使原平衡二叉树失去平衡。

此时将 A 的左孩子 B 提升为子树的根结点，A 作为 B 的右子树的根结点，而将 B 原来的右子树作为 A 的左子树。

（2）RR 型：新结点插入到 A 的右子树中而使原平衡二叉树失去平衡。

此时将 A 的右孩子 B 提升为子树的根结点，A 作为 B 的左子树的根结点，而将 B 原来的左子树作为 A 的右子树。

（3）LR 型：新结点插入到 A 的左孩子的右子树中而使原平衡二叉树失去平衡。

此时将 A 的左孩子 B 的右孩子 C 提升为子树的根结点，将 B、A 分别作为 C 的左、右子结点，而将 C 原来的左、右子树分别作为 B 的右子树和 A 的左子树。

（4）RL 型：新结点插入到 A 的右孩子的左子树中而使原平衡二叉树失去平衡。

此时将 A 的右孩子 B 的左孩子 C 提升为子树的根结点，将 A、B 分别作为 C 的左、右子结点，而将 C 原来的左、右子树分别作为 A 的右子树和 B 的左子树。

○平衡二叉树的查找分析：在查找过程中和给定值进行比较的关键字个数不超过树的深度，因此其平均查找的时间复杂度是 $O(\log n)$。

6.2.4 哈希表的查找

6.2.4.1 哈希表的概念

前面介绍的查找方法的共同特点在于：通过对关键字的一系列比较，逐步缩小查找范围，直到确定结点的存储位置或确定查找失败。而哈希法则希望不经过任何比较，一次存取就能得到所查元素，因此必须在记录的存储位置和它的关键字之间建立一个确定的对应关系，使得每个关键字和结构中一个唯一的存储位置相对应，这样查找时只需对结点的关键字进行某种运算就能确定结点在表中的位置。因此哈希法的平均比较次数和表中所含结点的个数无关。

基本思想：在记录的存储位置和它的关键字之间建立一个确定的对应关系 f，使每个关键字和表中一个惟一的存储位置相对应；这样，不经过比较，一次存取就能得到所查元素。称对应关系 f 为哈希（散列）函数。

哈希表：又叫散列表。应用哈希函数，由记录的关键字确定记录在表中的地址，并将记录存入此地址，这样构成的表称为哈希表。

哈希查找：又叫散列查找，利用哈希函数进行查找的过程称为哈希查找。

若 k1≠k2，而 f（k1）=f（k2），则这种现象称为冲突，且 k1 和 k2 对哈希函数 f 来说是同义词。

根据设定的哈希函数 f=H（key）和处理冲突的方法，将一组关键字映像到一个有限的连续的地址集上，并以关键字 key 在地址集中的"像"作为记录在表中的存储位置，这一映像过程称为构造哈希表。

哈希技术的应用关键在于解决两个问题：怎样找一个好的哈希函数；如何设计有效解决冲突的方法。

6.2.4.2 哈希函数的构造方法

一个好的哈希函数应该使一组关键字的哈希地址均匀地分布在整个哈希表中，从而减少冲突。但鉴于实际问题中关键字的不同，没法构造出统一的哈希函数，从而使构造哈希函数的方法多种多样，常用的构造哈希函数的方法有如下 6 种：

1）直接定址法（自身函数法）：取关键字或关键字的某个线性函数作哈希地址，也就是 H（key）=key 或 H（key）=a·key+b，其中 a 和 b 是常数。

特点：直接定址法所得地址集合与关键字集合大小相等，不会发生冲突。但这种函数只适用于给定的一组关键字为关键字集合中的全体元素的情况，故实际中能用这种哈希函数的情况很少。

2）数字分析法：对关键字进行分析，取关键字的若干位或其组合作哈希地址。

特点：适于关键字位数比哈希地址位数大，且可能出现的关键字事先知道的情况。

3）平方取中法：取关键字平方后的中间几位作为哈希地址，取的位数由表长决定。

特点：适于不知道全部关键字情况。

4）除留余数法：取关键字被某个不大于哈希表表长 m 的数 p 除后所得余数作哈希地址，即 H（key）=key MOD p，p≤m。

特点：简单、常用，可与上述几种方法结合使用。p 的选取很重要；p 选不好，容易产生同义词。一般可选 p 为素数。

5）随机数法：取关键字的随机函数值作哈希地址，即 H（key）=random（key）。

特点：适于关键字长度不等的情况。

6）折叠法：将关键字分割成位数相同的几部分，然后取这几部分的叠加和（舍去进位）作哈希地址。具体方法可分为两种：

○移位叠加，即将分割后的几部分低位对齐相加；

○间界叠加，即从一端沿分割界来回折叠，然后对齐相加。

特点：适于关键字位数很多，且每一位上数字分布大致均匀情况。

6.2.4.3 选取哈希函数的因素

计算哈希函数所需时间、关键字长度、哈希表长度（哈希地址范围）、关键字分布情况、记录的查找频率等。

6.2.4.4 哈希冲突的解决方法

均匀的哈希函数可以减少冲突但不能避免冲突，因此，必须有良好的方法来处理冲突。通常，处理冲突的方法有下列两种：

1. 开放定址法（线性探查法）

当冲突发生时，形成一个探查序列；沿此序列逐个地址探查，直到找到一个空位置（开放的地址），将发生冲突的记录放到该地址中，即 H_i=（H（key）+di） MOD m, i=1, 2, …, k（k≤m-1），其中：H（key）是哈希函数，m 为哈希表表长，di 为增量序列。

○具体方法分类

（1）线性探测法：di=1, 2, 3, …, m-1；

（2）平方探测法：di=1^2, -1^2, 2^2, -2^2, 3^2, …, $\pm k^2$（k≤m/2）；

（3）伪随机探测法：di=伪随机数序列；

（4）再哈希法（双哈希函数法）：构造第二个哈希函数，当发生冲突时，计算下一个哈希地址，即：di=i·RH（key），其中 RH 也是散列函数。

2. 链地址法

将所有关键字为同义词的记录存储在一个单链表中，并用一维数组存放头指针。具体方法是设立一个指针数组,初始状态是空指针,哈希地址为 i 的记录都插入到指针数组中第 i 项所指向的链表中,保持关键字有序。

6.3 实践案例

6.3.1 顺序查找

6.3.1.1 系统简介

依次输入顺序表中的各个元素,然后进行关键字查找。如果存在则返回待查元素的位置,否则显示不存在。

6.3.1.2 设计思路

○顺序查找的基本思想

对于给定的关键字 k,从表的一端开始,逐个进行数据元素的关键字和给定值的比较,若当前扫描到的结点关键字与 k 相等则查找成功;若扫描结束后,仍未找到关键字等于 k 的节点,则查找失败。

建立一个顺序表,数据元素从下标为 1 的单元开始放入,下标为 0 的单元起监视哨作用,将待查的关键字存入下标为 0 的单元,顺序表从后向前查找,若直到下标为 0 时才找到关键字则说明查找失败;若不到下标为 0 时就找到关键字,则查找成功。

6.3.1.3 程序清单

```
#include<stdio.h>
#define KEYTYPE int
#define MAXSIZE 100

/***************************************************************
    顺序表中的元素定义
***************************************************************/
typedef struct
{
    KEYTYPE   key;
}SSELEMENT;

/***************************************************************
    顺序表结构定义
***************************************************************/
typedef struct
{
    SSELEMENT r[MAXSIZE];
```

```c
    int len;
}SSTABLE;

/***********************************************************************
    顺序查找
***********************************************************************/
int seq_search(KEYTYPE k,SSTABLE *st)        //顺序表查找元素
{
    int j;
    j=st->len;                               //顺序表元素个数
    st->r[0].key=k;                          //st->r[0]单元作为监视哨
    while(st->r[j].key!=k)                   //顺序表从后向前查找
        j--;
    return j;                                //j=0,找不到,j≠0 找到
}

/***********************************************************************
    主函数
***********************************************************************/
main()
{
    SSTABLE   a;
    int i=0,j=0,k=1;
    printf("请输入顺序表元素,元素为整型量,用空格分开,-99 为结束标志：\n");

    scanf("%d",&i);
    while (i!=-99)
    {
        j++;
        a.r[k].key=i;
        k++;
        scanf("%d",&i);
    }
    a.len=j;

    while(i!=-1)
    {
        printf("\n 输入待查元素关键字： ");
        scanf("%d",&i);
        k=seq_search(i,&a);
```

```
            if(i==-1)  break;

            if(k==0)
            {
                printf("顺序表中待查元素不存在\n");
            }
            else
            {
                printf("顺序表中待查元素的位置是：%d",k);
                printf("\n");
            }
        }
}
```

6.3.1.4 运行结果

```
请输入顺序表元素,元素为整型量,用空格分开,-99为结束标志：
23 12 34 5 78 29 36 44 12 9
45 67 33 7 8 3 1 111 5 71
-99

输入待查元素关键字：12
顺序表中待查元素的位置是：9

输入待查元素关键字：34
顺序表中待查元素的位置是：3

输入待查元素关键字：2
顺序表中待查元素不存在

输入待查元素关键字：-1
Press any key to continue_
```

6.3.2 折半查找

6.3.2.1 系统简介

依次输入顺序表中的各个元素，然后进行关键字查找。如果存在则返回待查元素的位置，否则显示不存在。

6.3.2.2 设计思路

○折半查找的基本思想

设查找表中的元素存放在数组 r 中，数据元素的下标范围为[low，high]，要查找的关键字值为 key，中间元素的下标为 mid=（low+high）／2（向下取整），令 key 与 r[mid]的关键字比较：

若 key=r[mid].key，查找成功，下标为 m 的记录即为所求，返回 mid。

若 key<r[mid].key，所要找的记录只能在左半部分记录中，再对左半部分使用折半查找

若 key>r[mid].key，所要找的记录只能在右半部分记录中，再对右半部分使用折半查找法继续进行查找，搜索区间缩小了一半。

重复上述过程，直到找到查找表中某一个数据元素的关键字的值等于给定的值 key 说明查找成功；或者出现 low 的值大于 high 的情况，说明查找不成功。

建立一个有序表，数据元素从下标为 1 的单元开始放入。实现查找算法时，首先将 low 赋值为 1，high 等于最后一个数据元素的下标，然后将给定的关键字的值与查找区间[low, high]中间的数据元素的关键字比较， 实现查找过程。

6.3.2.3 程序清单

```c
#include<stdio.h>
#define KEYTYPE int
#define MAXSIZE 100


/************************************************************************
    顺序表中的元素定义
************************************************************************/
typedef struct
{
    KEYTYPE   key;
}SSELEMENT;


/************************************************************************
    顺序表结构定义
************************************************************************/
typedef struct
{
    SSELEMENT r[MAXSIZE];
    int len;
}SSTABLE;


/************************************************************************
    折半查找
************************************************************************/
int search_bin(KEYTYPE k,SSTABLE *st)
{                                        //有序表上折半查找
    int low,high,mid;
    low=1;                               //low=1 表示元素从下标为 1 的单元放起
    high=st->len;                        //high=st->len 表示最后一个元素的下标
    while(low<=high)                     //low<=high 为继续查找的条件
```

```c
    {
        mid=(low+high)/2;
        if(k==st->r[mid].key)                //k==st->r[mid].key 表示查找成功
            return mid;
        else if(k<st->r[mid].key)            //否则继续二分查找
            high=mid-1;
        else
            low=mid+1;
    }
    return 0;                                //查找不成功,返回 0
}

/*********************************************************************
       主函数
**********************************************************************/
main()
{
    SSTABLE   a;
    int i=0,j=0,k=1;
    printf("请输入顺序表元素,元素为整型量,用空格分开,-99 为结束标志：\n");

    scanf("%d",&i);
    while (i!=-99)
    {
        j++;
        a.r[k].key=i;
        k++;
        scanf("%d",&i);
    }
    a.len=j;

    while(i!=-1)
    {
        printf("\n 输入待查元素关键字：");
        scanf("%d",&i);
        k=search_bin(i,&a);
        if(i==-1) break;

        if(k==0)
        {
```

```
                printf("顺序表中待查元素不存在\n");
            }
            else
            {
                printf("顺序表中待查元素的位置是：%d",k);
                printf("\n");
            }
        }
    }
}
```

6.3.2.4 运行结果

```
请输入顺序表元素,元素为整型量,用空格分开,-99为结束标志:
1 3 5 7 9 11 13 15 17 19 21 23 25 27 29 -99
输入待查元素关键字: 5
顺序表中待查元素的位置是：3
输入待查元素关键字: 25
顺序表中待查元素的位置是：13
输入待查元素关键字: 35
顺序表中待查元素不存在
输入待查元素关键字: -1
Press any key to continue_
```

6.3.3 二叉排序树

6.3.3.1 系统简介

依次输入关键字并建立二叉排序树，实现二叉排序树的插入和查找功能。

6.3.3.2 设计思路

〇二叉排序树基本思想

二叉排序树就是指将原来已有的数据根据大小构成一棵二叉树，二叉树中的所有结点数据满足一定的大小关系，所有的左子树中的结点均比根结点小，所有的右子树的结点均比根结点大。

二叉排序树查找是指按照二叉排序树中结点的关系进行查找，查找关键字首先同根结点进行比较，如果相等则查找成功；如果比根结点小，则在左子树中查找；如果比根结点大，则在右子树中进行查找。这种查找方法可以快速缩小查找范围，大大减少查找关键字的比较次数，从而提高查找效率。

〇算法分析

算法的关键过程实际上就是二叉排序树的创建和查找两个步骤。

首先分析二叉排序树的创建运算，由于二叉排序树中的所有结点都有一个大小关系，因此，每个结点必须在二叉排序树中寻找其合适的位置。创建二叉排序树的第一步就是创建一个根结点，第二步就是将其他结点不断插入到二叉排序树中的合适位置。二叉排序树进行结点插入时，首先要为被插入结点寻找合适的插入位置。这个过程实际上就是一个关键值不断的比较的过程。

第一次比较是与二叉排序树的根结点比较，如果比根结点的关键值小，则插入到他的左子树中；如果比根结点的关键值大，则插入到它的右子树中。在子树中重复这样过程，直到找到合适的位置为止。

二叉排序树的查找运算实际上同二叉排序树的创建运算是一致的。区别在于，创建过程中要为二叉排序树中没有位置的关键字建立结点，而查找过程中，只是在二叉排序树中查找是否等于查找关键字值的结点存在，不管有还是没有，它都不会创建一个新的结点。

6.3.3.3 程序清单

```c
#include<stdio.h>
#include<malloc.h>
#define null 0
typedef int keytype;                    //查找关键字为整型数据

/***************************************************************************
        二叉排序树结点类型
***************************************************************************/
struct   Bsnode
{
    keytype data;                       //数据域
    struct Bsnode *Lchild;              //左孩子指针
    struct Bsnode *Rchild;              //右孩子指针
};

/***************************************************************************
        BSP 为二叉排序树结点类型指针
***************************************************************************/
typedef struct Bsnode *BSP;
typedef struct Bsnode BST;

/***************************************************************************
        为关键字 key 创建一个二叉排序树结点
***************************************************************************/
BSP createBst(keytype key)
{
    BSP s;
    s=(BSP)malloc(sizeof(BST));
    s->data=key;
    s->Lchild=s->Rchild=null;
    return(s);
```

}
/***
    将 S 指向的结点插入到 T 指向的二叉排序树中
***/
BSP   BSTinsert(BSP T,BSP S)
{
    BSP q,p;
    if(T==null)
        return(S);
    else
    {
        p=T;
        q=null;
        while(p)   //在二叉排序树中为 S 结点寻找插入位置
        {
            q=p;
            if(S->data==p->data)   //若 S 结点与二叉排序树中根结点的值相等,则不插入
            {
                printf("The input key(%d)iS have in!\n",S->data);
                return(T);
            }
            if(S->data < p->data)   //若比根结点值小,则在左子树中寻找插入位置
                p = p->Lchild;
            else   //否则在右子树中寻找插入位置
                p = p->Rchild;
        }
        if(S->data<q->data)
            q->Lchild=S;   //作为左孩子插入
        else
            q->Rchild=S;   //作为右孩子插入
        return(T);
    }
}

/***
    在二叉排序树 T 中查找是否存在关键字
***/
void search(BSP T,keytype x)
{

```c
    BSP p;
    if(T==null)
    {
        printf("error\n");
        return;
    }
    else
    {
        p=T;
            while(p)
            {
                if(p->data == x)        //如果根结点是查找的关键字，查找结束
                {
                    printf("search success!\n");
                    return;
                }
                else if(x < p->data)    //如果小于根结点，则在左子树中查找
                    p=p->Lchild;
                else
                    p=p->Rchild;        //否则,在右子树中查找
            }
            if(!p)
                printf("%d not exist!\n",x);
    }
}

/************************************************************************
    输出二叉排序树
************************************************************************/
void BSTPrint(BSP T)
{
    if(T)
    {
        BSTPrint(T->Lchild);
        printf("%d->",T->data);
        BSTPrint(T->Rchild);
    }
}
```

```c
/************************************************************************
    主函数
************************************************************************/
main()
{
    printf("\n**                    二叉排序树                      **\n");
    printf("=========================================================\n");
    printf("**              1 --- 建立二叉排序树                   **\n");
    printf("**              2 --- 插入元素                         **\n");
    printf("**              3 --- 查询元素                         **\n");
    printf("**              4 --- 退出程序                         **\n");
    printf("=========================================================\n");
    BSP T,S;
    keytype key;
    int select=0,flag=1;
    T=null;
    while(flag)                           //主菜单
    {
        printf("Please select:\t");

        scanf("%d",&select);
        switch(select)
        {
        case 1:                           //创建二叉排序树
            {
                printf("Please input key to create Bsort_Tree:\n");
                scanf("%d",&key);
                while( key!=0)
                {
                    S=createBst(key);
                    T=BSTinsert(T,S);
                    scanf("%d",&key);
                }
                printf("Bsort_Tree is:\n");
                BSTPrint(T);
                printf("\n\n");
                break;
            }
        case 2:                           //在二叉排序树中插入关键字
```

```
            {
                  printf("Please input key which inserted:     ");
                  scanf("%d",&key);
                  S=createBst(key);
                  T=BSTinsert(T,S);
                  printf("Bsort_Tree is:\n");
                  BSTPrint(T);
                  printf("\n\n");
                  break;
            }
      case 3:                              //在二叉排序树中查找关键字
            {
                  printf("Please input key which searched:    ");
                  scanf("%d",&key);
                  search(T,key);
                  printf("\n\n");
                        break;
            }
      case 4:                        //退出
            {
                  flag=0;
                  break;
            }
      default:
            {
                  printf("输入错误,请选择正确的操作! \n");
                  break;
            }

      }
   }
}
```

6.3.3.4 运行结果

```
==================================================
**            二叉排序树                      **
==================================================
**              1 --- 建立二叉排序树          **
**              2 --- 插入元素                **
**              3 --- 查询元素                **
**              4 --- 退出程序                **
==================================================
Please select: 1
Please input key to create Bsort_Tree:
12 34 67 48 19 44 21 30 19 7 4 24 9 88 100 100 0
The input key<19>iS have in!
The input key<100>iS have in!
Bsort_Tree is:
4->7->9->12->19->21->24->30->34->44->48->67->88->100->

Please select: 2
Please input key which inserted:   90
Bsort_Tree is:
4->7->9->12->19->21->24->30->34->44->48->67->88->90->100->

Please select: 3
Please input key which searched:   90
search success!

Please select: 3
Please input key which searched:   110
110 not exist!

Please select: 4
Press any key to continue_
```

6.3.4 哈希查找

6.3.4.1 系统简介

依次输入关键字并建立哈希表，进行关键字查找。如果存在则返回待查元素的位置，否则显示不存在。

6.3.4.2 设计思路

○哈希表查找基本思想

哈希表查找是一种基于尽可能不通过比较操作而直接得到记录的存储位置的想法而提出的一种特殊查找技术。它的基本思想是通过记录中的关键字的值 key 为自变量，通过某一种确定的函数 h，计算出函数值 h（key）作为存储地址，将相应关键字的记录存储到对应的位置上。而在查找时仍需要用这个确定的函数 h 进行计算，获得所要查找的关键字所在记录的存储位置。

除留余数法（Division Method）是用关键字 key 除以某个正整数 M，所得余数作为哈希地址的方法。对应的哈希函数 h（key）为 h（key）=key%M，一般情况下 M 的取值为不大于表长的质数。

用开放定址法解决冲突，形成下一个地址的形式是

$$Hi=（h（key）+di）\%M \quad i=1,2,\cdots,k（k \leq m-1）$$

式中，h（key）为哈希函数；M 为某个正整数；di 为增量序列；m 为表长。

线性探测再散列是将开放定址法中的增量序列 di 设定为从 1 开始一直到表长减 1 的整数序列：1，2，3，…，m-1（m 为表长）。

〇算法分析

算法的关键过程实际上就是 Hash 表的创建和查找两个步骤。

首先分析 Hash 表的创建运算：将由键盘输入的关键字 key 作为自变量，通过除留余数法构造哈希函数 h，计算出函数值 h（key）作为存储地址，将该关键字存储到对应的位置上。如果产生冲突，则采用线性探查法，从关键字的哈希地址开始向后扫描，直到找到空位置，将该关键字存储在这个位置，插入成功；若扫描到哈希表的最后仍没有找到空位置，则插入失败。

查找时仍利用除留余数法构造哈希函数 h，计算出函数值 h（key）获得所要查找的关键字所在记录的存储位置。若存储位置对应的数据元素的值与查找关键字相等，则查找成功，否则，采用线性探查法，从关键字的哈希地址开始向后扫描，直到找到与关键字相等的数据元素，查找成功；若扫描到哈希表的最后仍没有找到与关键字相等的数据元素，则查找失败，不存在与关键字相等的数据元素。

6.3.4.3 程序清单

```
#include<stdio.h>                  //哈希表长最大值
#define HM 20
#define M 19
#define FREE 0                     //空闲标记
#define SUCESS 1                   //成功
#define UNSUCESS 0                 //不成功
typedef int keytype;               // keytype 为整型

/******************************************************************
    哈希表结构定义
******************************************************************/
typedef struct                     // keytype 型关键字 key
{
    keytype key;
    int cn;                        //查找次数
}hashtable;

/******************************************************************
    哈希函数
******************************************************************/
int h(keytype key)
{
```

```c
        return(key%M);
}

/************************************************************************
    哈希表查找函数
*************************************************************************/
int HashSearch(hashtable ht[],keytype key)
{
    int d,i;
    i=0;
    d=h(key);                          //求哈希地址
    ht [d].cn=0;
    while((ht[d].key!=key)&&(ht[d].key!=FREE)&&(i<HM))
    {
        i++;                           //求下一个地址
        ht[d].cn++;                    //查找次数加 1
        d=(d+i)%M;                     //线性探查记录的插入位置
    }
    if(i>=HM)
    {
        printf("The hashtable is full!\n");
        return(UNSUCESS);
    }
    return(d);                         //若 h[d]的关键字等于 key 说明查找成功
}


/************************************************************************
    哈希表插入函数, 查找不成功的时候, 将给定关键字 key 插入哈希表中
*************************************************************************/
int HashInsert(hashtable ht[],keytype key)
{
    int d;
    d=HashSearch(ht,key);
    if(ht[d].key==FREE)
    {
        ht[d].key=key;
        return(SUCESS);                //插入成功
    }
    else
    {
```

```
        printf("Uncessful!\n");
        return(UNSUCESS);              // 插入不成功
    }
}

/**************************************************************************
    建立哈希表的函数
**************************************************************************/
void HashCreat(hashtable ht[])
{
    int i,n;
    keytype key1;
    printf("请输入元素个数(要小于表长%d)：\n",HM);
    scanf("%d",&n);
    printf("请输入元素关键字：\n");
    for(i=0;i<n;i++)
    {
        scanf("%d",&key1);
        HashInsert(ht,key1);           //调用哈希表插入算法
    }
}

/**************************************************************************
    主函数
**************************************************************************/
void main()
{
    hashtable ht[HM];                  //哈希表空间
    keytype key=0;
    int i;
    for(i=0;i<HM;i++)
    {
        ht[i].key=0;
    }
    printf("建立哈希表\n");
    HashCreat(ht);

    while(key!=-99)
    {
        printf("\n 输入待查元素关键字：");
```

```
            scanf("%d",&key);
            if(key==-1)    break;
            i=HashSearch(ht,key);

            if(ht[i].key==key){
                printf("哈希表中存在此元素，位置为%d\n",i);
            }
            else{
                printf("哈希表中无此元素\n");
            }
        }
    }
```

6.3.4.4 运行结果

```
建立哈希表
请输入元素个数<要小于表长20>：
10
请输入元素关键字：
11 15 18 36 9 55 7 4 1 80

输入待查元素关键字：15
哈希表中存在此元素，位置为15

输入待查元素关键字：55
哈希表中存在此元素，位置为1

输入待查元素关键字：1
哈希表中存在此元素，位置为2

输入待查元素关键字：100
哈希表中无此元素

输入待查元素关键字：-1
Press any key to continue_
```

6.4 巩固提高

编程实现一个开放式的高校本科招生最低录取分数线的查询系统，供师生和家长查询。高校自愿放入该校的信息，可能随时有高校加入。

○要求实现的查询功能有
（1）查询等于用户给定分数的高校；
（2）查询大于(或小于)用户给定分数的高校；
（3）查询最低录取分数线在用户给定的分数段中的高校。
○提示
该实验主要的功能是查找，查找表为高校最低录取分数信息的集合。根据题意可知，该查找表中的元素个数可能随时增减，所以它是一个动态查找表，可采用树状结构保存。为了提高查询速度，可建立二叉排序树并在二叉排序树中实现查找。

第一个要求可以直接利用一般的二叉排序树的查找算法实现。

第二个要求可以利用比较二叉排序树根结点的关键字值和给定的分数实现，若前者大于或等于后者，根结点及其右子树中的结点全部满足要求，再在根结点的左子树中查找；否则只在根结点的右子树中查找即可。

第三个要求可以利用比较二叉排序树根结点的关键字值和给定的分数段实现，若前者在给定的分数段中，根结点就满足要求，则在根结点的左、右子树中继续查找即可；若前者给定的分数段的左侧，只在根结点的右子树中查找即可；若前者在给定的分数段的右侧，只根节点的左子树中查找即可。

第七章 排序

排序是计算机程序设计中的一种重要操作,它的功能是将一个数据元素(或记录)的任意序列,重新排列成一个按关键字有序的序列。

从第六章的讨论中容易看出,人们通常希望计算机中的表是按关键字有序存储的。因为有序的表可以采用查找效率较高的折半查找法,其平均查找长度为 log2(n+1)-1,而无序的表只能进行顺序查找,其平均查找长度为(n+1)/2。又如建造树表(无论是二叉排序树或B-树)的过程本身就是一个排序的过程。因此,学习和研究各种排序方法是提高计算机应用效率的重要课题之一。

7.1 实践目的和要求

7.1.1 实践目的

1) 掌握各种排序算法的思想及其适用条件;
2) 掌握上机实现各种排序算法的基本方法;
3) 深刻理解排序的定义和各种排序方法的特点,并能加以灵活应用;
4) 掌握各种方法的排序过程及其依据的原则,并掌握各种排序方法的时间复杂性和稳定性的分析方法。

7.1.2 实践要求

1) 掌握本章实践案例程序的算法。
2) 上机运行本章的程序,保存和打印出程序的运行结果,并结合程序进行分析。
3) 比较各种排序算法的各自特点,能够根据实际情况选择合适的排序方法。

7.2 基本概念

7.2.1 排序的概念

为了便于查找,通常希望计算机中的数据表是按关键码有序的,因为使用有序表的折半查找效率较高。排序(Sorting)是计算机程序设计中的一种重要操作,其功能是对一个数据元素集合或序列重新排列成一个按关键字的值有序(本章假设为升序)的序列。

如果待排序的表中有多个关键字值相同的记录,经过排序后这些具有相同关键字值的记录之间的相对次序保持不变,则称此排序方法是稳定的;反之,若经过排序后这些具有相同关键字值的记录之间的相对次序发生了变化,则称这种排序方法是不稳定的。

在排序过程中，若待排序列的全部结点都存放在内存，并在内存中调整它们在待排序列中的顺序，称为内排序。在排序过程中，若待排序列只有部分结点被调入内存，并借助内存调整结点在外存中的存放顺序的排序方法称为外排序。外排序是对大型文件的排序，待排序的记录存储在外存中，在排序过程中，内存只存储文件的一部分记录，整个排序过程需进行多次内外存间的交换。

排序的定义：设有记录序列：{ R1, R2, …, Rn }，其相应的关键字序列为：{ K1, K2, …, Kn }。若存在一种确定的关系：$Kx \leq Ky \leq … \leq Kz$，则将记录序列 { R1, R2, …, Rn } 排成按该关键字有序的序列：{ Rx, Ry, …, Rz } 的操作，称为排序。

本章介绍的排序算法都是对顺序表进行内排序。

7.2.2 插入排序

插入排序的基本思想是：每次将一个待排序的记录，按其关键字值的大小插入到已经排序部分的文件中的适当位置上，直到全部插入完成。分为直接插入排序和堆排序两种方法。

7.2.2.1 直接插入排序

这是一种最简单的排序方法，其过程是依次将每个记录插入到一个有序的序列中去，从而得到一个新的、记录数增1的有序表。

○排序过程

在直接插入排序过程中，每将一个记录插入到前面已排好序的记录序列中，称为一趟直接插入排序。初始时，仅有一个记录的表是自然有序的，因此，对有 n 个记录的表，可从第二个记录开始直到第 n 个记录，逐个向有序表中进行插入操作，从而得到 n 个记录按关键码有序的表。

○效率分析

空间效率：仅用了一个辅助单元，所以空间复杂度是 $O(1)$。

时间效率：向有序表中逐个插入记录的操作，进行了 n-1 趟，每趟操作分为比较关键码和移动记录，而比较的次数和移动记录的次数取决于表的初始排列。

最好情况下：即表已有序，每趟操作只需 1 次比较 2 次移动。总比较次数=n-1 次，总移动次数=2（n-1）次。时间复杂度为 $O(n)$。

最坏情况下：即第 j 趟操作，插入记录需要同前面的 j 个记录进行 j 次关键码比较，移动记录的次数为 j+2 次。时间复杂度为 $O(n^2)$。

平均情况下：即第 j 趟操作，插入记录大约同前面的 j/2 个记录进行关键码比较，移动记录的次数为 j/2+2 次。时间复杂度为 $O(n^2)$。

由此，直接插入排序的时间复杂度为 $O(n^2)$。是一个稳定的排序方法。

对直接插入排序方法进行修改，将顺序查找待插入记录的位置改为折半查找，则得到折半插入排序方法。

7.2.2.2 希尔排序（缩小增量排序）

希尔（Shell）排序的基本思想是：把记录按下标的一定增量 d 进行分组，对每组记录采用直接插入排序方法进行排序，随着增量逐渐减小，所分成的组包含的记录越来越多，到增量的值减为 1 时，整个数据合成为一组，构成一个有序序列，则排序完成。

○排序过程

先取一个正整数 d1<n，把所有相隔 d1 的记录放一组，组内进行直接插入排序；然后取 d2<d1，重复上述分组和排序操作；直至 di=1，即所有记录放进一个组中排序为止。

○效率分析

对希尔排序进行效率分析很难，关键字的比较次数与记录移动次数依赖于增量序列的选取，特定情况下可以准确估算出关键字的比较次数和记录的移动次数。目前还没有人给出选取最好的增量序列的方法。增量序列可以有各种取法，有取奇数的，也有取质数的，但需要注意：增量中除 1 外没有公因子，且最后一个增量必须为 1。

空间效率：仅用了一个辅助单元，所以空间复杂度是 $O(1)$。

时间效率：平均情况下时间复杂度与最坏情况下时间复杂度均为 $O(nlog_2n)$。

希尔排序方法是一个不稳定的排序方法。

7.2.3 选择排序

选择排序的基本方法是：每一趟从待排序列中选取一个关键字值最小的记录，顺序放在已排序的记录序列的最后。也即第一趟从 n 个记录中选取关键字值最小的记录，第二趟从剩下的 n-1 个记录中选取关键字值最小的记录，直到整个序列的记录全部排完为止。

7.2.3.1 直接选择排序（简单选择排序）

○排序过程

第一趟，从 n 个记录中找出关键字值最小的记录与第一个记录交换；第二趟，从第二个记录开始的 n-1 个记录中再选出关键字值最小的记录与第二个记录交换；如此，第 i 趟，则从第 i 个记录开始的 n-i+1 个记录中选出关键字值最小的记录与第 i 个记录交换，重复上述操作，共进行 n-1 趟排序后，排序结束。

○效率分析

空间效率：仅用了一个辅助单元，所以空间复杂度是 $O(1)$。

时间效率：直接选择排序算法的效率与表的初始状态无关。平均情况下、最好情况下与最坏情况下时间复杂度均为 $O(n^2)$。

直接选择排序方法是一个不稳定的排序方法。

7.2.3.2 堆排序（Heap Sort）

○定义

n 个元素的序列（k1，k2，…，kn），把它的所有元素按完全二叉树的顺序存放在一个一维数组中，当且仅当满足下列关系之一时，称之为最小堆或者最大堆。

$$\begin{cases} k_i \leq k_{2i} \\ k_i \leq k_{2i+1} \end{cases} \quad 或 \quad \begin{cases} k_i \geq k_{2i} \\ k_i \geq k_{2i+1} \end{cases} \quad (i=1, 2, \cdots, \lfloor n/2 \rfloor)$$

○性质

由堆的定义可知，堆具有如下性质：

堆是一棵采用顺序存储结构的完全二叉树，k1 为根结点。

堆的根结点是元素序列中的值最小（此时的堆称为最小堆）或最大元素（此时的堆称为最大堆）。

从根结点到每一个叶子结点的路径上的元素组成的序列都是按元素值递增或递减的。

○堆排序的基本思想

首先，将无序序列 R[1, …, n]建成一个堆，交换 R[1]和 R[n]；然后，将 R[1, …, n-1]调整为堆，交换 R[1]和 R[n-1]；如此重复执行，直到交换了 R[1]和 R[2]为止，最终得到一个有序序列。

因此，实现堆排序需解决两个问题：

（1）建堆，即如何将 n 个元素的无序序列按关键字值建成堆；

（2）将剩余元素调整为堆，即将堆顶元素 R[1]与堆中最后一个元素 R[n]交换后，怎样调整剩余的前 n-1 个元素为一个新堆。

○建堆方法

建堆方法：对初始序列建堆的过程，就是一个反复进行筛选的过程。从无序序列的第 $\lfloor n/2 \rfloor$ 个元素（即此无序序列对应的完全二叉树的最后一个非终端结点）起，至第一个元素止，进行反复筛选。

将剩余元素调整为堆的方法：将根结点值与左、右子树的根结点值进行比较，并与其中小者进行交换；重复上述操作，直至叶子结点，将得到新的堆，称这个自根结点至叶子结点的调整过程为"筛选"。

○效率分析

空间效率：空间复杂度是 O(1)。

时间效率：平均情况下、最好情况下与最坏情况下时间复杂度均为 $O(n\log_2 n)$。

堆排序方法是一个不稳定的排序方法。

7.2.4 交换排序

交换排序的基本方法是：两两比较待排序记录的关键字值，并交换不满足次序要求的那些记录，直到全部满足为止。最常见的交换排序方法是冒泡排序和快速排序。

7.2.4.1 冒泡排序

○基本思想

通过无序区中相邻记录关键字值间的比较和位置的交换，使关键字值最小的记录如"气泡"一样逐渐漂浮到水面。

○排序过程

（1）将第一个记录 R[1]的关键字与第二个记录 R[2]的关键字进行比较，若为逆序 R[1].key>R[2].key，则交换 R[1]与 R[2]；然后比较第二个记录与第三个记录；依此类推，直至第 n-1 个记录和第 n 个记录比较完为止。这是第一趟冒泡排序，结果是关键字值最大的记录被放在最后 n 个记录上；

（2）对前 n-1 个记录进行第二趟冒泡排序，结果使关键字值次大的记录被放在第 n-1 个记录位置；

（3）重复上述过程，直到在某趟排序过程中没有进行过交换记录的操作为止，得到一个有序序列。

○效率分析

空间效率：空间复杂度是 O(1)。

时间效率：最好情况（有序序列）下只需 n-1 比较，时间复杂度为 O(n)，平均情况下与最坏情况（逆序序列）下时间复杂度均为 O(n^2)。

冒泡排序方法是一个稳定的排序方法。

7.2.4.2 快速排序（划分交换排序）

○基本思想

在待排序列中任取一个记录（通常取第一个记录）作为枢轴，通过一趟排序后，整个待排序列被枢轴记录分割成独立的两部分，其中一部分记录的关键字均比另一部分记录的关键字小，则可分别对这两部分记录进行快速排序，重复此过程直到整个序列有序。

○排序过程

对 R[s……t]中记录进行一趟快速排序，附设两个指针 i=s 和 j=s，设枢轴记录 rp=R[s]，x=rp.key。首先从 j 所指位置向前搜索第一个关键字小于 x 的记录，并和 rp 交换；再从 i 所指位置起向后搜索，找到第一个关键字大于 x 的记录，和 rp 交换；重复上述两步，直至 i==j 为止。第一趟快速排序结束。再分别对两个子序列进行快速排序，直到每个子序列只含有一个记录为止，得到一个有序序列。

○效率分析

空间效率：快速排序是递归的，每层递归调用时的指针和参数均要用栈来存放，递归调用层次数与 n 个记录对应的二叉树的深度一致。因而，存储开销在理想情况下为 O($\log_2 n$)，在最坏情况下即二叉树是一个单链时为 O(n)。

时间效率：最好情况（每次划分都正好将序列划分成两个等长的子序列）下时间复杂度为 O(n $\log_2 n$)，平均情况下与最坏情况（每次划分只得到一个子序列）下时间复杂度均为 O(n^2)。

快速排序通常被认为在同数量级（O(n$\log_2 n$)）的排序方法中平均性能最好的。但若初始序列有序或基本有序时，快排序反而蜕化为冒泡排序。为改进之，通常以"三者取中法"来选取枢轴记录，即取待排序列的两个端点与中点三个记录关键字值居中的记录作为枢轴。

快速排序是一个不稳定的排序方法。

7.2.5 归并排序

7.2.5.1 归并排序的概念

归并：将两个或两个以上的有序表组合成一个新的有序表，称为归并。

二路归并：将两个有序表合并为一个有序表。

7.2.5.2 二路归并排序的基本思想

只有一个元素的表总是有序的。所以对于有 n 个元素的待排序列，每个元素可看成一个长度为 1 有序子表。将相邻的有序表进行两两归并，得到 n/2 个长度为 2 的有序表；然后再将这些有序表两两归并，得到 n/4 个长度为 4 的有序表；如此反复，直到得到一个长度为 n 的有序表。

7.2.5.3 二路归并排序的效率分析

空间效率：归并过程中需要一个与表等长的辅助元素数组空间，所以空间复杂度为 O(n)。

时间效率：对于有 n 个元素的表，将这 n 个元素看作叶结点，若将两两归并生成的子表看作它们的父结点，则归并过程对应由叶向根生成一棵二叉树的过程。所以归并趟数约等于二叉树的高度-1，即 $\log_2 n$，每趟归并需移动记录 n 次，故时间复杂度为 $O(n\log_2 n)$。

归并排序是一个稳定的排序方法。

7.2.6 基数排序

7.2.6.1 基本定义和概念

基数排序是一种借助于多关键字排序的思想，将单关键字按基数分成"多关键字"进行排序的方法。

多关键字排序：设有 n 个记录的待排序列{R1, R2, …, Rn}且每个记录中包含 d 个关键字{k1, k2, …, kd}，则称序列{R1, R2, …, Rn}对关键字{k1, k2, …, kd}有序是指：对于序列中任两个记录 R[i]和 R[j]（1≤i≤j≤n）都满足下列有序关系：（$k1_i$, $k2_i$, …, kd_i）＜（$k1_j$, $k2_j$, …, kd_j）。其中 k1 称为最主位关键字，kd 称为最次位关键字。

多关键字排序按照从最主位关键字到最次位关键字或从最次位关键字到最主位关键字的顺序逐次排序，分两种方法：

（1）最高位优先法（MSD）：先对待排序列根据最高位关键字 k1 排序，将序列分成若干子序列，每个子序列有相同的 k1 值；然后让每个子序列对次关键字 k2 排序，又分成若干更小的子序列；依次重复，直至就每个子序列对最低位关键字 kd 排序；最后将所有子序列依次连接在一起成为一个有序序列。

（2）最低位优先法（LSD）：对待排序列从最低位关键字 kd 起进行排序，然后再对高一位的关键字排序；依次重复，直至对最高位关键字 k1 排序后，便得到一个有序序列。

基数排序：对于数字型或字符型的单关键字，可以看成是由多个数位或多个字符构成的多关键字，根据关键字中各位的值，通过对排序的 n 个元素进行若干趟"分配"和"收集"的办法进行排序，称作基数排序。

链式基数排序：用链表作存储结构的基数排序。

7.2.6.2 基数排序

设待排序的线性表中每个元素的关键字都是 d 位的十进制正整数。

在排序过程中需要对该线性表进行 d 趟的分配与收集处理，每趟处理方法是相同的。

在进行第 j（j=1, 2, 3, …, d）趟处理时，首先按元素在线性表中的排列顺序，依次将每个元素插入到编号为 0 或 9 的某个队列（关键字右起第 j 位上的值是几就插入到几号队列），这个过程叫作分配；然后，按队列编号从小到大、同一队列按插入先后的顺序，从队列中取出所有元素，重新构成一个线性表，这个过程叫作收集。

在进行了 d 趟分配与收集之后，排序过程结束，得到一个有序线性表。

7.2.6.3 链式基数排序过程

（1）设置 10 个队列，f[i]和 e[i]分别为第 i 个队列的头指针和尾指针；
（2）分配时，按当前关键字位所取值，改变记录的指针值，将记录分配到 10 个不同的链队列中，每个队列中记录的关键字位的值相同；
（3）收集时，改变所有非空队列的队尾记录的指针域，令其指向下一个非空队列的队头记录,重新将 10 个队列链成一个链表即按当前关键字位取值从小到大将各队列首尾相连成一个链表；
（4）对每个关键字位重复（2）、（3），最后得到一个有序序列。

7.2.6.4 基数排序的效率分析

空间效率：需要 2*r（基数）个指向队列的辅助空间，以及用于链表的 n 个指针。所以空间复杂度为 O(n+r)。

时间效率：设待排序列为 n 个记录，d 个关键字位，关键字位的取值范围为 r，则进行链式基数排序的时间复杂度为 O(d(n+r))，其中，一趟分配时间复杂度为 O(n)，一趟收集时间复杂度为 O(r)，共进行 d 趟分配和收集。

基数排序是一个稳定的排序方法。

7.2.7 各种排序方法比较

下面通过一个表格简单比较一下几种常见的内排序方法之间的性能特点（表 7.1）。

表 7.1 排序方法及性能汇总表

排序方法		平均时间	最坏情况	最佳情况	辅助空间	是否稳定
简单排序	选择排序	$O(n^2)$	$O(n^2)$	$O(n^2)$	$O(1)$	不稳定
	直接插入排序	$O(n^2)$	$O(n^2)$	$O(n)$	$O(1)$	稳定
	冒泡排序	$O(n^2)$	$O(n^2)$	$O(n)$	$O(1)$	稳定
希尔排序		$O(n\log_2 n)$	$O(n^2)$	—	$O(1)$	不稳定
快速排序		$O(n\log_2 n)$	$O(n^2)$	$O(n\log_2 n)$	$O(\log_2 n)$	不稳定
堆排序		$O(n\log_2 n)$	$O(n\log_2 n)$	$O(n\log_2 n)$	$O(1)$	不稳定
归并排序		$O(n\log_2 n)$	$O(n\log_2 n)$	$O(n\log_2 n)$	$O(n)$	稳定
基数排序		$O(d(n+rd))$	$O(d(n+rd))$	$O(d(n+rd))$	$O(r+n)$	稳定

7.3 实践案例

7.3.1 系统简介（8 种排序算法比较案例）

随机函数产生 10000 个随机数，用快速排序、直接插入排序、冒泡排序、选择排序的排序方法排序，并统计每一种排序所花费的排序时间和交换次数。其中随机数的个数由用户定义，系统产生随机数。

7.3.2 设计思路

为了产生随机数，必须用到头文件 stdlib.h 中的两个函数 srand()和 rand()来设置随机种子以及产生随机数。为了能够计时，必须用到头文件 time.h 中的 time()和 difftime()两个函数，time()用于截取计算机内的时钟，difftime()用于得到两次时钟间隔的时间（秒）。每一种排序方法都单独写成子函数形式，然后用主函数调用它。为了能查看排序前后的效果，可以单独写一个子函数输出数组结果，在排序前后分别调用它，可以看到排序前后的结果。

7.3.3 程序清单

```
#include "iostream.h"
#include "stdio.h"
#include "stdlib.h"
#include "time.h"

/***************************************************************************
    冒泡排序
***************************************************************************/
long Bubblesort(long R[], long n)
{
    int flag=1;                        //当 flag 为 0，则停止排序
    long BC=0;
    for(long i=1;i<n;i++)
    {                                  //i 表示趟数，最多 n-1 趟
        flag=0;                        //开始时元素未交换
        for(long j=n-1;j>=i;j--)
        {
            if(R[j]<R[j-1])            //发生逆序
            {
                long t=R[j];
                R[j]=R[j-1];
                R[j-1]=t;flag=1;       //交换，并标记发生了交换
            }
            BC++;
        }
    }
    return BC;
}
```

```
/***************************************************************************
    选择排序
***************************************************************************/
long selectsort(long R[], long n)
{
    long i,j,m;long t,SC=0;
    for(i=0;i<n-1;i++)
    {
        m=i;
        for(j=i+1;j<n;j++)
        {
            SC++;
            if(R[j]<R[m]) m=j;
            if(m!=i)
            {
                t=R[i];
                R[i]=R[m];
                R[m]=t;
            }
        }
    }
    return SC;
}


/***************************************************************************
    直接插入排序
***************************************************************************/
long insertsort(long R[], long n)
{
    long IC=0;
    for(long i=1;i<n;i++)                    //i 表示插入次数,共进行 n-1 次插入
    {
        long temp=R[i];                      //把待排序元素赋给 temp
        long j=i-1;
        while((j>=0)&&(temp<R[j]))
        {
            R[j+1]=R[j];j--;                 //顺序比较和移动
            IC++;
```

```
            }
            IC++;
            R[j+1]=temp;
        }
        return IC;
}


/***************************************************************
    希尔排序
***************************************************************/
long ShellSort(long R[], int n)
{
    int temp,SC=0;
    for(int i = n / 2; i > 0; i /= 2)              //将所有记录分成增量为 t 的子序列
    {
        for(int j = 0; j < i; j ++)                //对每个子序列进行插入排序
            for(int k = j + i; k < n; k += i)      //依次将记录插入有序子序列中
                for(int p = j; p < k; p += i)      //循环查找要插入的位置
                    if (R[k] < R[p]) {
                        temp = R[k];
                        for(int q = k; q > p; q -= i){   //插入位置以后的记录依次后移
                            R[q] = R[q - i];
                            SC++;
                        }
                        R[p] = temp;               //插入记录
                        break;
                    }
    }
    return SC;
}


/***************************************************************
    快速排序
***************************************************************/
long quicksort(long R[], long left, long right)
{
    static long QC=0;
    long i=left,j=right;
    long temp=R[i];
    while(i<j)
```

```
            {
                while((R[j]>temp)&&(j>i))
                {
                    QC++;
                    j=j-1;
                }
                if(j>i)
                {
                    R[i]=R[j];
                    i=i+1;
                    QC++;
                }
                while((R[i]<=temp)&&(j>i))
                {
                    QC++;
                    i=i+1;
                }
                if(i<j)
                {
                    R[j]=R[i];
                    j=j-1;
                    QC++;
                }
            }
                                            //二次划分得到基准值的正确位置
        R[i]=temp;
        if(left<i-1)
            quicksort(R,left,i-1);          //递归调用左子区间
        if(i+1<right)
            quicksort(R,i+1,right);         //递归调用右子区间
        return QC;
}


/*****************************************************************
    堆排序
*****************************************************************/
static long HC=0;
void Heap(long R[], int n)                  //重新构造小顶堆
{
```

188

```
    int temp;
    for(int i = 0; i * 2 < n; i ++)
    {
        if (R[i] >= R[2 * i] && R[2 * i]) {
            temp = R[i];
            R[i] = R[2 * i];
            R[2 * i] = temp;
            HC++;
        }
        if (R[i] >= R[2 * i + 1] && R[2 * i + 1]) {
            temp = R[i];
            R[i] = R[2 * i + 1];
            R[2 * i + 1] = temp;
            HC++;
        }
    }
}

long HeapSort(long R[], int n)              //取出堆顶
{
    for(int i = n - 1; i >= 0; i --)
    {
        Heap(R, i);
        R[0] = R[i];
    }
    return HC;
}


/*********************************************************************
    归并排序
*********************************************************************/
static long MC=0;
void Merge(long c[], long d[], int l, int m, int r)
{//合并c[1:m]和c[m+1:r]到d[1:r]
    int i = l, j = m + 1, k = l;
    while ((i <= m) && (j <= r)) {
        if (c[i] <= c[j])
            d[k ++] = c[i ++];
        else
```

```cpp
                d[k ++] = c[j ++];
            MC++;
        }
        if (i > m)
            for(int q = j; q <= r; q ++)
                d[k ++] = c[q];
        else
            for(int q = i; q <= m; q ++)
                d[k ++] = c [q];
}

void Copy(long a[], long b[], int left, int right)
{
    for(int i = left; i <= right; i ++)
        a[i] = b[i];
}

long MergeSort(long a[], int left, int right)         //递归将数组分成子数组
{
    long* b = new long [right + 1];
    if (left < right) {
        int i = (left + right) / 2;
        MergeSort(a, left, i);
        MergeSort(a, i + 1, right);
        Merge(a, b, left, i, right);
        Copy(a, b, left, right);
    }
    return MC;
}

/************************************************************************
    基数排序
************************************************************************/
class node                                            //结点类
{
    friend class list;
public:
    int value;
    node* next;
    node(){value = 0; next = NULL;}
```

```cpp
protected:
private:
};

class list                              //链表类
{
public:
    void Insert(list& L, int value);
    node* head;
    list(){head = NULL;}
protected:
private:
};

void list::Insert(list &L, int value)   //插入结点
{
    node* N = new node;
    N->value = value;
    if (!L.head) {
        L.head = N;
    }
    else {
        node* p = L.head;
        while (p->next) {
            p = p->next;
        }
        p->next = N;
    }
}

long RadixSort(long R[], int n)
{
    list a[10];
    int i,RC=0;;
    for(int t = 1; t < 1000; t *= 10)
    {
        for(i = 0; i < 10; i ++)
        {
            int j = R[i] / t % 10;
            a[j].Insert(a[j], R[i]);
```

```
            }
            int n = 0;
            for(i = 0; i < 10; i ++)
            {
                node* p = a[i].head;
                while (p) {
                    R[n] = p->value;
                    p = p->next;
                    n ++;
                }
                a[i].head = NULL;
            }
        }
        return RC;
}


/**************************************************************************
    操作选择函数
**************************************************************************/
void operate(long a[], long n)
{
    long * R = new long [n];
    time_t start, end;
    double dif;
    long degree;
    char ch;
    cout << "请选择排序算法：\t";
    cin >> ch;
    switch(ch){
    case '1':
        {
            for(int i = 0; i < n; i ++)
            {
                R[i] = a[i];
            }
            time(&start);
            degree = Bubblesort(R, n);
            time(&end);
```

```cpp
                dif = difftime(end, start);
                cout << "冒泡排序所用时间：\t" << dif << '\n';
                cout << "冒泡排序交换次数：\t" << degree << '\n';
                cout << '\n';
                operate(a, n);
                break;
            }
        case '2':
            {
                for(int i = 0; i < n; i ++)
                {
                    R[i] = a[i];
                }
                time(&start);
                degree = selectsort(R, n);
                time(&end);
                dif = difftime(end, start);
                cout << "选择排序所用时间：\t" << dif << '\n';
                cout << "选择排序交换次数：\t" << degree << '\n';
                cout << '\n';
                operate(a, n);
                break;
            }
        case '3':
            {
                for(int i = 0; i < n; i ++)
                {
                    R[i] = a[i];
                }
                time(&start);
                degree = insertsort(R, n);
                time(&end);
                dif = difftime(end, start);
                cout << "直接插入排序所用时间：   " << dif << '\n';
                cout << "直接插入排序交换次数：   " << degree << '\n';
                cout << '\n';
                operate(a, n);
                break;
            }
        case '4':
```

```cpp
        {
            for(int i = 0; i < n; i ++)
            {
                R[i] = a[i];
            }
            time(&start);
            degree = ShellSort(R, n);
            time(&end);
            dif = difftime(end, start);
            cout << "希尔插入排序所用时间：   " << dif << '\n';
            cout << "希尔插入排序交换次数：   " << degree << '\n';
            cout << '\n';
            operate(a, n);
            break;
        }
    case '5':
        {
            for(int i = 0; i < n; i ++)
            {
                R[i] = a[i];
            }
            time(&start);
            degree = quicksort(R, 0, n - 1);
            time(&end);
            dif = difftime(end, start);
            cout << "快速排序所用时间：\t" << dif << '\n';
            cout << "快速排序交换次数：\t" << degree << '\n';
            cout << '\n';
            operate(a, n);
            break;
        }
    case '6':
        {
            for(int i = 0; i < n; i ++)
            {
                R[i] = a[i];
            }
            time(&start);
            degree = HeapSort(R, n);
            time(&end);
```

```cpp
                dif = difftime(end, start);
                cout << "堆排序所用时间：\t" << dif << '\n';
                cout << "堆排序交换次数：\t" << degree << '\n';
                cout << '\n';
                operate(a, n);
                break;
            }
        case '7':
            {
                for(int i = 0; i < n; i ++)
                {
                    R[i] = a[i];
                }
                time(&start);
                degree = MergeSort(R, 0, n);
                time(&end);
                dif = difftime(end, start);
                cout << "归并排序所用时间：\t" << dif << '\n';
                cout << "归并排序比较次数：\t" << degree << '\n';
                cout << '\n';
                operate(a, n);
                break;
            }
        case '8':
            {
                for(int i = 0; i < n; i ++)
                {
                    R[i] = a[i];
                }
                time(&start);
                degree = RadixSort(R, n);
                time(&end);
                dif = difftime(end, start);
                cout << "基数排序所用时间：\t" << dif << '\n';
                cout << "基数排序交换次数：\t" << degree << '\n';
                cout << '\n';
                operate(a, n);
                break;
            }
        case '9':
```

```cpp
                break;
        default:
            {
                cout << "输入错误，请选择正确的操作！" << '\n';
                break;
            }
        }

}

/*************************************************************************
        主函数
*************************************************************************/
void main()
{
    cout<<"\n**                  排序算法比较                    **"<<endl;
    cout<<"===================================================="<<endl;
    cout<<"**            1 --- 冒泡排序                        **"<<endl;
    cout<<"**            2 --- 选择排序                        **"<<endl;
    cout<<"**            3 --- 直接插入排序                    **"<<endl;
    cout<<"**            4 --- 希尔排序                        **"<<endl;
    cout<<"**            5 --- 快速排序                        **"<<endl;
    cout<<"**            6 --- 堆排序                          **"<<endl;
    cout<<"**            7 --- 归并排序                        **"<<endl;
    cout<<"**            8 --- 基数排序                        **"<<endl;
    cout<<"**            9 --- 退出程序                        **"<<endl;
    cout<<"===================================================="<<endl;

    cout << "\n 请输入要产生的随机数的个数：";
    long n;
    cin >> n;
    cout << endl;
    long *a = new long [n];
    srand((unsigned long)time(NULL));
    for (long i = 0; i < n; i ++)
    {
        a[i] = rand() % n;
    }

    operate(a, n);
}
```

7.3.4 运行结果

```
**              排序算法比较              **
**                                        **
**        1 —— 冒泡排序                  **
**        2 —— 选择排序                  **
**        3 —— 直接插入排序              **
**        4 —— 希尔排序                  **
**        5 —— 快速排序                  **
**        6 —— 堆排序                    **
**        7 —— 归并排序                  **
**        8 —— 基数排序                  **
**        9 —— 退出程序                  **

请输入要产生的随机数的个数：10000

请选择排序算法：      1
冒泡排序所用时间：    1
冒泡排序交换次数：    49995000

请选择排序算法：      2
选择排序所用时间：    1
选择排序交换次数：    49995000

请选择排序算法：      3
直接插入排序所用时间：0
直接插入排序交换次数：24952382

请选择排序算法：      4
希尔插入排序所用时间：1
希尔插入排序交换次数：151833

请选择排序算法：      5
快速排序所用时间：    0
快速排序交换次数：    155612

请选择排序算法：      6
堆排序所用时间：      1
堆排序交换次数：      21287965

请选择排序算法：      7
归并排序所用时间：    1
归并排序比较次数：    120415

请选择排序算法：      8
基数排序所用时间：    0
基数排序交换次数：    0

请选择排序算法：      9
Press any key to continue
```

7.4 巩固提高

设计一个学生成绩管理系统，对某个班级的学生的 5 门功课的学习成绩进行管理。
○要求
（1）求每门功课的平均成绩。
（2）输出每门课程成绩优秀的学生名单及成绩。
（3）输出只要有一门课程不及格的学生名单及其每门成绩。
（4）对 5 门课程中可以指定某一门课程进行排序。

○提示

（1）由于要对学生成绩进行排序，难免要进行数据移动，考虑采用何种存储方式可以提高算法效率。

（2）每个学生有 5 门成绩，考虑用何种存储方式较为合适。

（3）考虑用何种排序方法实现较适合。

（4）如果数据量较大，采用什么方法能使排序效率提高。

第八章 综合篇

至此我们已经学习了《数据结构》课程中的主要数据结构及相关算法,本章通过两个综合性的项目来讨论各种类型的数据结构的基本概念、逻辑结构、存储结构以及相关算法的实现及应用,并为各种现有存储方式的应用设计相应的算法,以期培养学生运用数据结构相关知识解决实际问题的能力。

8.1 目的和要求

8.1.1 实践目的

1)提高综合运用各种数据结构及其算法解决应用问题的能力。
2)扩大编程量,完成模块化程序设计的全过程。

8.1.2 实践要求

1)掌握本章案例的算法。
2)上机运行案例程序,保存和打印出程序的运行结果,并结合程序进行分析。
3)按照你的操作需要,重新改写程序并运行,打印出文件清单和运行结果。

8.2 相关概念

在迷宫益智游戏中采用了栈和队列两种数据结构,分别对应于算法一和算法二。

算法一采用了回溯法,回溯法也称为试探法,该方法首先暂时放弃关于问题规模大小的限制,并将问题的候选解按某种顺序逐一枚举和检验。当发现当前候选解不可能是解时,就选择下一个候选解;倘若当前候选解除了还不满足问题规模要求外,满足所有其他要求时,继续扩大当前候选解的规模,并继续试探。如果当前候选解满足包括问题规模在内的所有要求时,该候选解就是问题的一个解。在回溯法中,放弃当前候选解,寻找下一个候选解的过程称为回溯。扩大当前候选解的规模,以继续试探的过程称为向前试探。在用回溯法求解问题时一般采用的是将解空间组织成状态空间树并进行深度优先遍历的方法。在遍历过程中,先遍历的结点后回溯,而后遍历的结点先回溯,即被遍历的结点符合栈的"后进先出"特点,所以采用了栈这种数据结构。这时,不仅可以用栈来表示正在遍历的树的结点,而且可以很方便地表示建立孩子结点和回溯过程。

算法二采用了分支限界算法,分支限界法是一种用于求解组合优化问题的排除非解的搜索算法。类似于回溯法,分支定界法在搜索解空间时,也经常使用树形结构来组织解空间。然而与回溯法不同的是,分支限界一般用广度优先方法来搜索这些树。在遍历树的过程当中,先遍历的结点先扩展，而后遍历的结点后扩展，即被遍历的结点符合队列的"先进先出"特点，所以采用了队列这种数据结构。

在景区旅游信息管理系统中采用了图这种数据结构，因为旅游信息管理系统中的景点是以图形的形式分布的，在程序设计中采用图作为数据结构来存储景点信息是很自然的。

8.3 实践案例

8.3.1 迷宫益智游戏

8.3.1.1 系统简介

迷宫的结构布局可自定义设置，骑士从迷宫入口走进迷宫，迷宫中设置很多墙壁，对前进方向形成了多处障碍。骑士需要在迷宫中寻找通路以到达出口。

8.3.1.2 设计流程

将迷宫的结构布局设计成由起点（红旗，深色）、终点（蓝旗，灰色）、围墙（最外一圈）、障碍（黑色）和可通过点（空白处，表示可通过）这五部分组成，其中，迷宫起点、终点和障碍物可由玩家自定义设置。

在搜索迷宫时，从起点出发，若遇到围墙或障碍则表示不能通过。只有是当下一个路径是可通过的时候，才能继续往下搜索，否则尝试另一个方向。

迷宫的具体结构布局如下图所示：

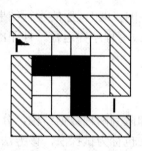

图 8.1 迷宫结构示意图

本系统使用了两种算法，实现了路径试探查询和最快路径查询这两种路径查询方法，包括了对四方向和八方向两种路径探寻方向。

两种算法描述如下：

○算法一

为了搜索一条通往终点的路径，我们从起点开始出发，对途中经过的每一位置的四个方向（上下左右）进行探索，如果有一个方向可通，即往那个方向前进一步。然后搜索新位置的各个方向，直到搜索到终点为止。若一个位置的4个方向都不通，则返回这个位置的上一个方向，搜索上一个位置的下一个方向的路径。

迷宫的求解过程可以采用回溯法，即在一定的约束条件下试探地搜索前进，若前进中受阻（碰上围墙或障碍或是已经走过的路……），则返回并向下一个方向进行新的搜索。

○ 算法二

为了能够达到能一次性搜索到一条最快的路径而不用回溯,可以采用另一种方法:分支限界法。分支限界法的意思是对当前结点,先从左到右产生它的所有儿子,按约束函数检查,只要有一个儿子不超越约束条件,它就是可活动的结点,将它加入活动结点表尾。以后从活动结点表中依次取出一个结点作为当前扩展结点,并产生它的所有子结点,将其加入活动结点表的末尾,产生顺序仍然是从左至右。

假设以起点为中心,它旁边的 4 个方向的位置(上,下,左,右)距离起点的距离为 1。这些与起点相距为 1 的位置的旁边那些位置(不包括已经是与起点相距为 1 的位置)与起点的距离就是 2,并且 2 是这些点到起点的最短距离。以此类推下去,与起点相距为 N 的位置,它们的旁边的位置的点(不包括已经设置了与起点距离的位置)与起点的距离为 N+1,并且 N+1 是这些点到起点的最短距离。

假设探索到终点这个位置到起点的距离为 N,那么终点旁边一定有一个 N-1 的位置,因为 N 是从 N-1 这个点 + 1 才得到的。那么从 N 这个点到 N-1 这个点是从起点到终点最短距离的一部分。然后从 N-1 到 N-2,……,2 到 1,1 到起点,这些位置连起点就是一条从起点到终点的最快路径。

具体搜索的示意图如图 8.2 和图 8.3 所示。

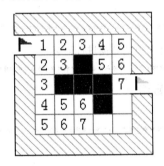

图 8.2 广度搜索过程

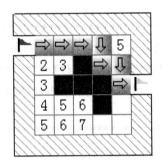

图 8.3 由图 8.2 得到的最终路径

8.3.1.3 数据结构

○ 算法一

算法一在求解过程采用了回溯法,运用堆栈就成了一个很自然的想法。

1) 运用堆栈进行搜索时,按照一个既定的方向顺序进行探索,每探索到一个可通过点,则将该点的位置压入堆栈,直到探索到终点位置结束搜索。

2) 如果在搜索中发现该方向不通,即尝试下一个方向。

3) 如果所有的方向都不通,则表示这个位置是一个不可通过点,可以为该点设置一个"脚印"标志,然后将该位置出栈,返回到上一个找到的位置进行它的下一个方向的探索。

4) 最终当堆栈为空(也就是说连起点也出栈了),就表明该迷宫不能出去。

5) 当在探索过程中发现了终点,就表明搜索到了一个出路。

○ 算法二

算法二运用分支限界算法的时候,采用了队列的数据结构和广度优先的算法。

1) 在探索过程中,为了能找到最快的出路,我们从起点开始,将起点周围的可通过点作为活动结点进队,并计算出这些活动结点距起点的距离,然后将起点出队,表示起点周围的位置都已标识完毕。

2) 然后依次从队列中取出一个位置作为当前扩展结点,并产生它的所有子结点(也就是它旁边的那些可通过点),将将这些子结点当作活动结点入队,并计算出这些活动结点距起点的最近距离(也就是当前扩展结点的距离加1)。最后将该扩展结点出队。

3) 最终当队列为空了就表明该迷宫不能出去。

4) 当在搜索过程中搜索到一终点的时候,将终点到起点的距离记录下来,开始往回搜索最快的路径。每次将比该保存的距离小1的位置即为最短路径中的一个位置,直到回到起点为止即找到了一条最短的路径。

○程序设计和头文件定义

```
/************************************************************************
    迷宫的位置坐标在程序中用 CPoint 来代表
    CPoint::x 表示位置的横坐标,CPoint::y 表示位置的纵坐标。
************************************************************************/
        CPoint      m_ptStart;                  //保存起点坐标
        CPoint      m_ptEnd;                    //保存终点坐标


/************************************************************************
    为了便于对迷宫进行操作,将迷宫声明为一个二维数组,对迷宫里的数据进行记录。
************************************************************************/
        int**       m_pMazeData;                //迷宫二维数组


/************************************************************************
    布局数据变量
************************************************************************/
        int         m_nXStartPos;               // 迷宫 X 方向起点
        int         m_nYStartPos;               // 迷宫 Y 方向起点
        int         m_nXEndPos;                 // 终点
        int         m_nYEndPos;                 // 终点

        int         m_nXViewPort;               // 视图原点
        int         m_nYViewPort;               // 当视图滚动时触发 OnDraw()

        int         m_nRows;                    // 迷宫的行数
        int         m_nColumns;                 // 迷宫的列数

        int         m_nCellSize;                // 单位格大小

/************************************************************************
    枚举表示 m_pMazeData[y][x] 的值的具体意义
************************************************************************/
```

```
enum
{
    MAZE_NORMAL = -1,
    MAZE_SELECTED = -2,
    MAZE_START    = -3,           //    起点
    MAZE_END      = -4,           //    终点
    MAZE_WALL     = -5,           //    外围围墙
    MAZE_STONE    = -6,           //    表示内部障碍——石头
    MAZE_FOOTMARK = -7,           //    足迹,"脚印"

    // 表示这个点暂时还没有下一步（也没有方向），另有用法为 MAZE_PATH + 方向值 = 该类型
    MAZE_PATH         = 0,

    // 迷宫的路径为了在绘制的时候方便,被分为 8 个方向来表示,分别表示从当前位置要到下一步的方向
    MAZE_PATH_TOP        = 1,
    MAZE_PATH_RIGHTTOP   = 2,
    MAZE_PATH_RIGHT      = 3,
    MAZE_PATH_RIGHTDOWN  = 4,
    MAZE_PATH_BOTTOM     = 5,
    MAZE_PATH_LEFTBOTTOM = 6,
    MAZE_PATH_LEFT       = 7,
    MAZE_PATH_LEFT_TOP   = 8
};

/***************************************************************************
    八个用来搜索的方向
***************************************************************************/
enum
{
    MAZE_TOP         = 1,
    MAZE_RIGHTTOP    = 2,
    MAZE_RIGHT       = 3,
    MAZE_RIGHTBOTTOM = 4,
    MAZE_BOTTOM      = 5,
    MAZE_LEFTBOTTOM  = 6,
    MAZE_LEFT        = 7,
    MAZE_LEFTTOP     = 8
};
```

```cpp
/***************************************************************
        算法类型变量
****************************************************************/
        int         m_iAlgorithmType;           //记录使用何种算法
    enum
        {
            MAZE_ALGORITHM_STACK   =1,
            MAZE_ALGORITHM_QUEUE   =2
        };
/***************************************************************
        本系统采用了 STL 中的 stack 和 queue 容器，容器中存储了搜索过程中的路径位置。
****************************************************************/
        stack<CPoint>     m_stackMaze;
        queue<CPoint>     m_queueMaze;
/***************************************************************
        系统设置变量
****************************************************************/
        BOOL        m_b4Direction;          // 是4方向迷宫还是8方向迷宫
        // 是否显示搜索过程，如果显示，在搜索过程中将向视图类发送数据
        BOOL        m_bShowProcess;
        int         m_iProcessSpeed;        // 在演示过程中要等待的时间（ms）
/***************************************************************
        给界面发送的消息类别
****************************************************************/
        #define WM_MAZEPATH_PROCESS     10010       //发送  搜索过程
        #define WM_MAZEPATH_FIND        10011       //发送  搜索完毕
        #define WM_MAZEPATH_NORESULT    10012       //发送  没有路可达终点
/***************************************************************
        自定义消息宏
****************************************************************/
        #define WM_SETTING_UPDATE           10001   //更新迷宫
        #define WM_DEFINE_MAZE              10002   //定义迷宫
        #define WM_SETMAZE_DTYPE_4          10003   //四方向寻路
        #define WM_SETMAZE_DTYPE_8          10004   //八方向寻路
        #define WM_SET_SHOWPROCESS          10005   //显示搜索过程
        #define WM_SET_PROCESS_SPEED        10006   //设置演示速度
        #define WM_SETMAZE_ATYPE_STACK      10007   //堆栈算法实现
        #define WM_SETMAZE_ATYPE_QUEUE      10008   //队列算法实现
```

8.3.1.4 程序清单

系统的树形类结构如下图所示：

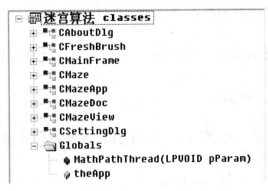

图 8.4　树形类结构图

其中，CMainFrame 类用来显示系统框架，CMazeApp 类是系统的应用程序类，CMazeView 类显示迷宫，CMazeDoc 类保存迷宫数据，CSettingDlg 类实现设置对话框的相关操作，MathPathThread 为全局的线程函数，CFreshBrush 类用来设置画刷属性，其中包含了一个动态链接库（#pragma comment(lib,"MyFreshBrush.lib")）。

CMaze 类封装了一个迷宫类 CMaze，实现迷宫的各种操作。

/***
　　[作用]　　搜索函数调用接口
　　[原理]　　该成员函数依据当前的算法类型变量调用相应的方法
　　[返回值]　返回值为 true 时为找到路径, false 时为没有路。
***/
bool CMaze::MazePath()
{
if(this->m_iAlgorithmType == CMaze::MAZE_ALGORITHM_STACK)
　　return MazePath_Stack();　　　　　　　//调用算法一
else if(this->m_iAlgorithmType == CMaze::MAZE_ALGORITHM_QUEUE)
　　return MazePath_Queue();　　　　　　　//调用算法二
return false;
}

○算法一——堆栈迷宫程序
/***
　　[作用]　　MazePath_Stack 实现了算法一
　　[原理]　　主要通过调用 getNextValidCell()函数实现回溯法搜索路径
***/
bool CMaze::MazePath_Stack()
{

```cpp
    int size = int(this->m_stackMaze.size());              // 清空操作
    for(int i = 0 ; i<size; i++)
         this->m_stackMaze.pop();
    ASSERT(m_stackMaze.empty());


    this->m_stackMaze.push(this->m_ptStart);               // 放入第一个

    while(!m_stackMaze.empty())
    {
        /* 获取当前位置，及其通往下一步的方向，如果是 0，则表明没有下一步，或者是
        下一步没有路走，退回 */
        CPoint pCur = m_stackMaze.top();
        int    nCurD = this->GetCellType(pCur.y,pCur.x) ;

        /* 获取它下一步该走的方向，如果是第一步，则起点为-3，如果是刚发现的点，则
        表明没有下一步（0）*/
        if(nCurD <= 0)           //
             nCurD = MAZE_TOP;
        else
        {
             if(this->m_b4Direction)
                  nCurD+=2;
             else
                  nCurD++;
        }

        // 获取下一步能走的坐标
        CPoint oldCurPoint = pCur;
        int nRet = this->getNextValidCell(pCur.y,pCur.x,nCurD);
        if( nRet == 1)
        {
            // 返回下一步可走的位置坐标和通向这个坐标的位置
            // 保存当前这个坐标的方位走向
            if(oldCurPoint != this->m_ptStart)
                this->SetMazeCellType(oldCurPoint.y,oldCurPoint.x,CMaze::MAZE_PATH + nCurD);

            // 压栈新位置信息
            this->m_stackMaze.push(pCur);

            // 如果到达了终点
```

```cpp
            if(pCur == this->m_ptEnd)
            {
                //这个应该是在 nRet = 2 里面处理
                ASSERT(FALSE);
            }
            // 将新找到的这个点设置为0方向,表示 它还没有下一步.
            this->SetMazeCellType(pCur.y,pCur.x,CMaze::MAZE_PATH);

            // 演示过程的话,就给视图发送数据,如果有演示速度要求的话,暂停...
            if(m_bShowProcess)
            {
    ::SendMessage(m_hViewWnd,WM_MAZEPATH_PROCESS,WPARAM(&pCur),nCurD);
                if(this->m_iProcessSpeed != 0)
                    ::Sleep(m_iProcessSpeed);
            }
        }
        else if( nRet == 2)    //到达终点
        {
            //AfxMessageBox(_T(" find one path! "));
            if(oldCurPoint != this->m_ptStart)
                this->SetMazeCellType(oldCurPoint.y,oldCurPoint.x,CMaze::MAZE_PATH + nCurD);
            ::SendMessage(m_hViewWnd,WM_MAZEPATH_FIND,0,0);
            return true;
        }
        else
        {
            // 说明当前位置已经没有路可走了, 设置为 FOOTMARK, 表示已经走过了但走不通
            if(this->GetCellType(pCur.y,pCur.x) == CMaze::MAZE_START)
            {
                AfxMessageBox(_T(" There is no path to the End! "));
                ::SendMessage(m_hViewWnd,WM_MAZEPATH_NORESULT,0,0);
                return false;
            }
            ASSERT(this->GetCellType(pCur.y,pCur.x) >= CMaze::MAZE_PATH);
            this->SetMazeCellType(pCur.y,pCur.x,CMaze::MAZE_FOOTMARK);

            // 出栈
            this->m_stackMaze.pop();
```

```cpp
            if(m_bShowProcess)
            {
                ::SendMessage(m_hViewWnd,WM_MAZEPATH_PROCESS,WPARAM(&pCur),-1);
                if(this->m_iProcessSpeed != 0)
                    ::Sleep(m_iProcessSpeed);
            }
        }
    }
    return false;
}

/***************************************************************************
    [作用]    getNextValidCell 用于堆栈搜索路径
    [原理]    对当前点(nCol, nRow)的 nextDirection 这个方向进行搜索，如果这个方向没有路径，则继续
搜索下一个方向（nextDirection++）
    [返回值]  如果找到终点，则函数返回 2，如果找到一个方向的可通过点，则返回这个可通过点（nCol,
nRow）和从原来那个点到现在这个可通过点的方向(nextDirection)
****************************************************************************/
int CMaze::getNextValidCell(LONG& nRow, LONG& nCol, int& nextDirection)
{
    if(nextDirection > 8)
        return 0;

    LONG tempRow = nRow, tempCol = nCol;
    switch(nextDirection)
    {
    case CMaze::MAZE_TOP:
        tempRow --;
        break;
    case CMaze::MAZE_RIGHTTOP:
        tempRow --;
        tempCol ++;
        break;
    case CMaze::MAZE_RIGHT:
        tempCol ++;
        break;
    case CMaze::MAZE_RIGHTBOTTOM:
        tempRow ++;
        tempCol ++;
```

```cpp
            break;
        case CMaze::MAZE_BOTTOM:
            tempRow++;
            break;
        case CMaze::MAZE_LEFTBOTTOM:
            tempRow ++;
            tempCol --;
            break;
        case CMaze::MAZE_LEFT:
            tempCol --;
            break;
        case CMaze::MAZE_LEFTTOP:
            tempRow --;
            tempCol --;
            break;
        default:
            ASSERT(FALSE);    //这个应该被 if(nextDireciton > 8 )处理
    }
    // 先判断是否出界
    if(tempRow >= 0 && tempCol >= 0 && tempRow < this->m_nRows && tempCol < this->m_nColumns)
    {
        // 再判断是否正常点
        if(this->GetCellType(tempRow,tempCol) == CMaze::MAZE_NORMAL)
        {
            nRow = tempRow;
            nCol = tempCol;
            return 1;
        }
        else if(this->GetCellType(tempRow,tempCol) == CMaze::MAZE_END)
        {
            //AfxMessageBox(_T("In getnextvalidcell find maze_end"));
            nRow = tempRow;
            nCol = tempCol;
            return 2;
        }
    }
    // 这个方向不通，尝试下一个方向
    if(this->m_b4Direction)
        nextDirection += 2;
```

```cpp
        else
            nextDirection ++;
        return getNextValidCell(nRow,nCol,nextDirection);


    return false;
}
```

○算法二——队列迷宫程序

```
/*************************************************************************
    [作用]    MazePath_Queue 实现了算法二
    [原理]    主要通过调用 setNeighborValue()和 getPrevNeighbor()函数实现分支限界法搜索路径
*************************************************************************/
bool CMaze::MazePath_Queue()
{
    // 先清空队列
    int nSize =int( m_queueMaze.size());
    for(int i = 0 ; i<nSize; i++)
        m_queueMaze.pop();
    ASSERT(m_queueMaze.empty());

    // 起点 进队列
    this->m_queueMaze.push(this->m_ptStart);

    int nEndValue = 0; // 保存最后的结果距离
    while(!m_queueMaze.empty())
    {
        // 1，取得队首者
        CPoint pCur = m_queueMaze.front();
        int cellValue = 10;
        if(pCur != m_ptStart)
            cellValue = this->GetCellType(pCur.y,pCur.x);
            cellValue -= 10;
        // 2, 将 pCur 的四周坐标点加入队列，并设置它们的值为当前点的值+1
        if( setNeighborValue(pCur.y,pCur.x,cellValue  + 1) )
        {
            // 返回到了终点, 退出循环去回溯
            nEndValue = cellValue   + 1;
            break;
        }
```

```cpp
        else
        {
            // 没找到终点，将当前这个点出队
            this->m_queueMaze.pop();
        }
    }
    if(m_queueMaze.empty())   // 说明是因为没有路了才退出的循环
    {
        AfxMessageBox(_T(" There is no path to the End! "));
        ::SendMessage(m_hViewWnd,WM_MAZEPATH_NORESULT,0,0);
        return false;
    }
    else       // 从终点开始寻找回起点的路
    {
        CPoint pCur = this->m_ptEnd;
            // 比如起点和终点之间的距离为 2，中间只隔了一个
            // 那么 nLastValue = 2，现在要循环 1 次就行，
            // 只要去指把中间的那个设置为相应的方向。
        while( 1 )
        {
            // 返回上一个点的坐标，从上一个点通往这个点的方向.
            int nDirection = getPrevNeighbor(pCur.y,pCur.x,--nEndValue);
            if(nDirection == 0) //找回到了起点
            {
                ::SendMessage(m_hViewWnd,WM_MAZEPATH_FIND,0,0);
                break;
            }
            this->SetMazeCellType(pCur.y,pCur.x,CMaze::MAZE_PATH + nDirection);
            if(m_bShowProcess)
            {
::SendMessage(m_hViewWnd,WM_MAZEPATH_PROCESS,WPARAM(&pCur),nDirection);
                if(this->m_iProcessSpeed != 0)
                    ::Sleep(m_iProcessSpeed);
            }
        }
    }
    return false;
}
```

/***
　　[作用]　　setNeighborValue 用来在广度搜索过程中，对当前位置(nCol, nRow)的附近位置设置它们距离起点的距离（nCurValue，当前位置距起点距离为 nCurValue-1）

　　[原理]　　当前位置的附近位置必须满足：
如果该位置不在迷宫之中，不设置。
如果该位置是终点位置，那么函数返回 true，表示找到终点。
如果该位置不是可通过点，不设置。包括起点、终点、围墙、障碍、已设置过距离的点。
如果当前位置是可通过点，那么将这个点的值（m_pMazeData[nRow][nCol]）设置为 nCurValue + 10。在这里 +10 主要是为了避免当前点距起点距离和另外一个类成员变量 MAZE_PATH_TOP ~ MAZE_PATH_LEFT_TOP 想混淆。因为最终迷宫的路径上的点的值为被设置为方向类型 MAZE_PATH_TOP ~ MAZE_PATH_LEFT_TOP，而不是距离。这是为了后面视图绘制迷宫时的方便。

　　[返回值]　　如果找到终点返回 true，否则返回 false
***/
bool CMaze::setNeighborValue(int nRow,int nCol,int　nCurValue)
{
　　// 1，先从方向 "上" 开始
　　int　nDirection = MAZE_TOP;
　　// 2，遍历
　　int tempRow,tempCol;
　　while(nDirection <= 8)
　　{
　　　　tempRow = nRow;
　　　　tempCol　　= nCol;
　　　　switch(nDirection)
　　　　{
　　　　case CMaze::MAZE_TOP:
　　　　　　tempRow --;
　　　　　　break;
　　　　case CMaze::MAZE_RIGHTTOP:
　　　　　　tempRow --;
　　　　　　tempCol ++;
　　　　　　break;
　　　　case CMaze::MAZE_RIGHT:
　　　　　　tempCol ++;
　　　　　　break;
　　　　case CMaze::MAZE_RIGHTBOTTOM:
　　　　　　tempRow ++;
　　　　　　tempCol ++;
　　　　　　break;
　　　　case CMaze::MAZE_BOTTOM:

```cpp
            tempRow++;
            break;
        case CMaze::MAZE_LEFTBOTTOM:
            tempRow ++;
            tempCol --;
            break;
        case CMaze::MAZE_LEFT:
            tempCol --;
            break;
        case CMaze::MAZE_LEFTTOP:
            tempRow --;
            tempCol --;
            break;
        default:
            ASSERT(FALSE);                      // 这个应该被 while(nDirection <= 8)处理
        }
        if(this->m_b4Direction)
            nDirection += 2;
        else
            nDirection ++;
                                                // 如果出界了
        if(tempRow < 0 || tempRow >= this->m_nRows
            || tempCol < 0 || tempCol >= this->m_nColumns )
        {
            continue;
        }
                                                // 获取周围一个点的类型
        int nType =  GetCellType(tempRow,tempCol);
        if(nType == MAZE_END)                   // 如果是终点
        {
            return true;
        }
                                                // 如果不是空点，进行下一次设置
        if (nType != MAZE_NORMAL)
            continue;
        else
        {
            CPoint temp(tempCol,tempRow);       // 入队
            this->m_queueMaze.push(temp);       // 设置 Value
```

// 为了与方向类型区别开来，在设置活动点类型的时候，将其值加上 10，
// 在用的时候再减 10
SetMazeCellType(tempRow,tempCol, nCurValue + 10);   //发送视图更新信息

if(m_bShowProcess)
{
    ::SendMessage(m_hViewWnd,WM_MAZEPATH_PROCESS,
                          WPARAM(&temp),nCurValue + 10);
    if(this->m_iProcessSpeed != 0)
        ::Sleep(m_iProcessSpeed);
}
}
}
return false;
}

/***
    [作用]   getPrevNeighbor 用来在 setNeighborValue 找到了终点位置后，用于往回找出最短路径中的前一个位置。
    [原理]   给定当前位置（nCol,nRow），以及要找的当前位置的附近位置的值必须是 nLastValue+10
    [返回值] 如果找到起点返回 0，否则返回前一个点的坐标，以及从前一个点通往当前这个点的方向
***/
int   CMaze::getPrevNeighbor(long&   nRow,long&   nCol, int   nLastValue)
{
    int   nDirection = MAZE_TOP;
    bool   bFind = false;
    // 这里的方向是从上一个点到这个点的方向。
    int tempRow = 0,tempCol = 0;
    while(nDirection <= 8)
    {
        tempRow = nRow;
        tempCol = nCol;
        switch(nDirection)
        {
        case CMaze::MAZE_TOP:
            tempRow ++;
            break;
        case CMaze::MAZE_RIGHTTOP:
            tempRow ++;
            tempCol --;

```cpp
        break;
    case CMaze::MAZE_RIGHT:
        tempCol --;
        break;
    case CMaze::MAZE_RIGHTBOTTOM:
        tempRow --;
        tempCol --;
        break;
    case CMaze::MAZE_BOTTOM:
        tempRow --;
        break;
    case CMaze::MAZE_LEFTBOTTOM:
        tempRow --;
        tempCol ++;
        break;
    case CMaze::MAZE_LEFT:
        tempCol ++;
        break;
    case CMaze::MAZE_LEFTTOP:
        tempRow ++;
        tempCol ++;
        break;
    default:
        ASSERT(FALSE); //这个应该被 while(nDirection <= 8)处理
    }

    if(tempRow < 0 || tempRow >= this->m_nRows
        || tempCol < 0 || tempCol >= this->m_nColumns )
    {
        nDirection += 2;
        if(nDirection > 8 && this->m_b4Direction == FALSE)
        {
            nDirection = 2;
        }
        continue;
    }
    int nType = GetCellType(tempRow,tempCol) ;
    if(nType == MAZE_START)
    {
        //AfxMessageBox(_T("finished!"));
```

```
            return 0;
        }
        if( nType    == nLastValue + 10)
        {       //    说明找到了上一个
            nRow = tempRow;
            nCol = tempCol;

            bFind = true;
            return    nDirection;
        }
        else
        {
            nDirection += 2;
            if(nDirection > 8 && this->m_b4Direction == FALSE)
            {
                nDirection = 2;
            }
        }
    }
    ASSERT(bFind);
    return 0;
}


/********************************************************************
    [作用]    为了在进行迷宫搜索的时候,不影响其他操作,在此为迷宫搜索开辟一个线程。也可以进行
挂起和恢复及结束这个线程。
********************************************************************/
UINT __cdecl MathPathThread( LPVOID pParam )
{
    CMaze *pMaze =(CMaze*) pParam;
    pMaze->MazePath();
    return 0;
}


/********************************************************************
    [作用]    用多线程开启迷宫搜索。
********************************************************************/
void CMazeView::OnMenuStart()
{
    if(this->m_pMazePathThread == NULL)
```

```cpp
        m_pMazePathThread=::AfxBeginThread( MathPathThread,GetDocument()->m_pMaze);
    else
        m_pMazePathThread->ResumeThread();
}


/************************************************************************
    [作用]    暂停搜索
************************************************************************/
void CMazeView::OnMenuPause()
{
    this->m_pMazePathThread->SuspendThread();
}


/************************************************************************
    [作用]    结束线程
************************************************************************/
void    CMazeView::OnMenuStop()
{
    if(this->m_pMazePathThread != NULL )
    {
        ::TerminateThread(this->m_pMazePathThread->m_hThread,0);
        this->m_pMazePathThread = NULL;
    }
}



/************************************************************************
    [作用]    迷宫界面的初始化
************************************************************************/
void CMazeView::OnInitialUpdate()
{

    CScrollView::OnInitialUpdate();

    SetScrollSizes(MM_TEXT, CSize( ::GetSystemMetrics(SM_CXSCREEN),
                                    ::GetSystemMetrics(SM_CYSCREEN)));

    if(this->m_bFirstInitUpdate) // 第一次打开程序
    {
        this->m_pMainFrame = (CMainFrame*)AfxGetMainWnd();
```

```cpp
    this->m_pSetDlg = new CSettingDlg();
    this->m_pSetDlg->Create(m_pSetDlg->IDD);
    this->m_pSetDlg->SetOwner(this);
    this->m_pSetDlg->CenterWindow(this);

    // 初使化一个渐变画刷
    this->m_brSelected.SetBrushSize(this->m_nCellSize,this->m_nCellSize);
    this->m_brSelected .CreateDIBBrush(RGB(255,255,255),RGB(0,255,0),TRUE,1);

    this->m_brMazePathV.SetBrushSize(this->m_nCellSize,this->m_nCellSize);
    this->m_brMazePathV.CreateDIBBrush(RGB(255,255,255),RGB(0,128,255),FALSE,1);
    this->m_brMazePathH.SetBrushSize(this->m_nCellSize,this->m_nCellSize);
    this->m_brMazePathH.CreateDIBBrush(RGB(255,255,255),RGB(0,128,255),TRUE,1);

    this->m_whiteBrush.CreateStockObject(WHITE_BRUSH);
    this->m_blackBrush.CreateStockObject(BLACK_BRUSH);
    this->m_hatchBrush.CreateHatchBrush(HS_FDIAGONAL ,RGB(128,120,120));

    m_fontMazePath.CreatePointFont(120,_T("宋体"));
    // 加载图片
    ASSERT(this->
            m_imageList.Create(IDB_BITMAP_MAZECELL,20,20,RGB(192,192,192)));

    // 抽取出里面的 Icon
    this->m_hIconStart = m_imageList.ExtractIcon(0);
    this->m_hIconEnd = m_imageList.ExtractIcon(1);
    this->m_hIconMazeFootMark    = m_imageList.ExtractIcon(2);

    this->GetDocument()->
            m_pMaze = new CMaze(this->m_hWnd,this->m_nRows,this->m_nColumns);
    this->m_bFirstInitUpdate = false;
}
else // 通过打开文件而调用这里的
{
    this->m_brSelected.ResetBrushSize(this->m_nCellSize,this->m_nCellSize);
    this->m_brMazePathH.ResetBrushSize(this->m_nCellSize,this->m_nCellSize);
    this->m_brMazePathV.ResetBrushSize(this->m_nCellSize,this->m_nCellSize);

    // 解决在找开新文件后，开始按钮不出现的情况
```

```cpp
        this->m_pMainFrame->
                m_wndToolBar.GetToolBarCtrl().HideButton(ID_START,FALSE);
        this->m_pMainFrame->RecalcLayout();
        // 重新打开一个文件后,设置滚动栏数据
        CRect rect;
        this->GetClientRect(&rect);

        SCROLLINFO si;
        si.cbSize   = sizeof(SCROLLINFO);
        si.nMin     =   0;
        si.nPos     =   0;
        si.nTrackPos = 0;
        si.nPage =rect.Width() ;
        si.nMax = m_nXEndPos + 50;

        si.fMask = SIF_ALL;
        this->SetScrollInfo(SB_HORZ,&si,0);
        si.nMax = m_nYEndPos+50;
        si.nPage = rect.Height();
        this->SetScrollInfo(SB_VERT,&si);
    }
}


/************************************************************************
    [作用]   绘制迷宫框架
************************************************************************/
void CMazeView::OnDraw(CDC* pDC)
{
    //先获取这个时候的 CScrollView 的视图焦点
    CPoint p = pDC->GetViewportOrg();
    this->m_nXViewPort = p.x;
    this->m_nYViewPort = p.y;

    CPen roundPen(PS_SOLID,2,COLORREF(RGB(0,0,0)));
    CPen *pOldPen = pDC->SelectObject(&roundPen);

    //   绘制边框
    pDC->MoveTo(this->m_nXStartPos,this->m_nYStartPos);
    pDC->LineTo(m_nXEndPos + 1,this->m_nYStartPos);
    pDC->LineTo(m_nXEndPos + 1,m_nYEndPos + 1);
```

```
pDC->LineTo(this->m_nXStartPos,m_nYEndPos + 1);
pDC->LineTo(this->m_nXStartPos,this->m_nYStartPos);
roundPen.DeleteObject();
// 绘制迷宫内列线
CPen innerPen(PS_SOLID,1,RGB(0,128,128));
pDC->SelectObject(&innerPen);
for(int i = 1; i<this->m_nColumns ; i++)
{
    pDC->MoveTo(this->m_nXStartPos + i * this->m_nCellSize, this->m_nYStartPos);
    pDC->LineTo(this->m_nXStartPos + i * this->m_nCellSize, m_nYEndPos);
}
for(i = 1; i<this->m_nRows ; i++)
{
    pDC->MoveTo(this->m_nXStartPos , this->m_nYStartPos + i * this->m_nCellSize);
    pDC->LineTo(m_nXEndPos,this->m_nYStartPos + i * this->m_nCellSize);
}

pDC->SelectObject(pOldPen);
roundPen.DeleteObject();
innerPen.DeleteObject();

// 在周围画上障碍
// 上下两边
SetBrushOrgEx(pDC->m_hDC,this->m_nXStartPos + this->m_nXViewPort,
    this->m_nYStartPos + this->m_nYViewPort,NULL);

CBrush brush;
brush.CreateHatchBrush(HS_FDIAGONAL ,RGB(128,120,120));
CRect rect(this->m_nXStartPos,this->m_nYStartPos,
        this->m_nXEndPos + 1,this->m_nYStartPos + this->m_nCellSize);
pDC->FillRect(&rect,&brush);
//rect.MoveToY(this->m_nYEndPos - this->m_nCellSize + 1);
rect.top = this->m_nYEndPos - this->m_nCellSize + 1;
rect.bottom = rect.top + m_nCellSize;
pDC->FillRect(&rect,&brush);
// 左右两边
rect.right = this->m_nXStartPos + this->m_nCellSize;
rect.top   = this->m_nYStartPos + this->m_nCellSize;
rect.bottom = this->m_nYEndPos - this->m_nCellSize + 1;
pDC->FillRect(&rect,&brush);
```

```cpp
    //rect.MoveToX(this->m_nXEndPos - this->m_nCellSize+1 );
    rect.left = this->m_nXEndPos - this->m_nCellSize+1;
    rect.right = rect.left + m_nCellSize;
    pDC->FillRect(&rect,&brush);
    brush.DeleteObject();

    // 画上终点和起点
    this->DrawACell(this->GetDocument()->m_pMaze->m_ptStart.x,
            this->GetDocument()->m_pMaze->m_ptStart.y,CMaze::MAZE_START,pDC);
    this->DrawACell(this->GetDocument()->m_pMaze->m_ptEnd.x,
            this->GetDocument()->m_pMaze->m_ptEnd.y,CMaze::MAZE_END,pDC);

    // CELL 内部画上
    for(i = 1;i<m_nRows -1;i++)
        for(int j = 1; j< m_nColumns -1 ; j ++)
        {
            // 不画 NORMAL 类型 CELL，加快效率
            int nStyle = this->GetDocument()->m_pMaze->GetCellType(i,j);
            if(nStyle != CMaze::MAZE_NORMAL)
                this->DrawACell(j,i,nStyle,pDC);
        }
}


/******************************************************************
    [作用]    利用鼠标单击操作设置迷宫的障碍
******************************************************************/
void CMazeView::OnLButtonDown(UINT nFlags, CPoint point)
{
    this->SetCapture();
    this->m_bMouseDown = true;

    switch(this->m_nCurrentStatus)
    {
    case STATUS_DEFINE_MAZE_START:
        {
            if(this->m_nXOldFocusCell != -1 && this->m_nYOldFocusCell != -1)//要在围墙上
            {
                // 设置迷宫里的障碍
                CPoint oldStartPt = this->GetDocument()->m_pMaze->m_ptStart;
                if(this->GetDocument()->m_pMaze->
```

```cpp
                SetMazeCellType(m_nYOldFocusCell,m_nXOldFocusCell,
                CMaze::MAZE_START))
            {
                CRect rect(CPoint(
                    this->m_nXStartPos + oldStartPt.x * this->m_nCellSize,
                    this->m_nYStartPos + oldStartPt.y * this->m_nCellSize),
                    CSize(m_nCellSize,m_nCellSize));
                //将这个范围扩大一些，这样就能完全重画那个 CELL；
                //rect.DeflateRect(-2,-2,-2,-2);
                rect.OffsetRect(this->m_nXViewPort,this->m_nYViewPort);

                /*在这里省略判断 rect 是否在 Client 区域里面，如果不在的话那肯定
                就不用画了，不过也不是说效率就会高出很多。*/
                this->InvalidateRect(&rect,1);;
                this->UpdateWindow();
                this->GetDocument()->SetModifiedFlag();
            }
        }
    }
    break;
case STATUS_DEFINE_MAZE_END:
    {
        if(this->m_nXOldFocusCell != -1 && this->m_nYOldFocusCell != -1)
        {
            //   设置迷宫里的障碍
            CPoint oldEndPt = this->GetDocument()->m_pMaze->m_ptEnd;
            if(this->GetDocument()->m_pMaze->SetMazeCellType(m_nYOldFocusCell,
                            m_nXOldFocusCell, CMaze::MAZE_END)) {
                CRect rect(CPoint(
                    this->m_nXStartPos + oldEndPt.x * this->m_nCellSize,
                    this->m_nYStartPos + oldEndPt.y * this->m_nCellSize),
                    CSize(m_nCellSize,m_nCellSize));
                //将这个范围扩大一些，这样就能完全重画那个 CELL；
                //rect.DeflateRect(-2,-2,-2,-2);
                rect.OffsetRect(this->m_nXViewPort,this->m_nYViewPort);

                /*在这里省略判断 rect 是否在 Client 区域里面*/
                this->InvalidateRect(&rect,1);
                this->UpdateWindow();
                this->GetDocument()->SetModifiedFlag();
```

```
                    }
                }
            }
            break;
        case STATUS_DEFINE_MAZE_STONE:
            {
                if(this->m_nXOldFocusCell != -1 && this->m_nYOldFocusCell != -1)
                {
                    if(this->GetDocument()->m_pMaze->SetMazeCellType(m_nYOldFocusCell,
                            m_nXOldFocusCell, CMaze::MAZE_STONE) {
this->DrawACell(m_nXOldFocusCell,m_nYOldFocusCell,CMaze::MAZE_STONE);
                    }
                    else
this->DrawACell(m_nXOldFocusCell,m_nYOldFocusCell,CMaze::MAZE_NORMAL);
                    this->GetDocument()->SetModifiedFlag();
                }
            }
            break;
    }
    CScrollView::OnLButtonDown(nFlags, point);
}

/*************************************************************************
    [作用]    绘制鼠标移动
    [原理]    绘制时,先用上一次的选择范围,当鼠标滚动时,通过 point 偏移来实现转换
*************************************************************************/
void CMazeView::OnMouseMove(UINT nFlags, CPoint point)
{
    // 检查 point 的范围
    CPoint mousePoint(point);           // 在绘制 STONE 时有用,先保存一个
    point.Offset(-1*this->m_nXViewPort,-1* this->m_nYViewPort);
    CRect mazeRect(m_nXStartPos,m_nYStartPos,m_nXEndPos,m_nYEndPos);
    int x , y ;
    if(mazeRect.PtInRect(point))
    {
        // 检查 point 的具体位置
        x = (point.x - this->m_nXStartPos)/this->m_nCellSize ;
        y = (point.y - this->m_nYStartPos)/this->m_nCellSize ;
    }
    else
```

```cpp
    {   x = -1;    y = -1;    }    // 不在 MAZE 里
m_pMainFrame->SetPaneOfRolcol(y,x);
// 和上一次位置相同则不变
if(x == this->m_nXOldFocusCell && y == this->m_nYOldFocusCell)
{
    CScrollView::OnMouseMove(nFlags, point);
    return;
}

// 将上一次的旧区域再画一次，如果上一次在外部，则不用画
if(this->m_nXOldFocusCell!= -1 && this->m_nYOldFocusCell!= -1)
{
            this->DrawACell(m_nXOldFocusCell,m_nYOldFocusCell,
this->GetDocument()->m_pMaze->GetCellType(m_nYOldFocusCell,m_nXOldFocusCell));
}
if(x != -1 && y != -1)
// 绘制新区域，设置高亮范围，即鼠标移动到的那一块上显示高亮
{
    switch(this->m_nCurrentStatus)
    {
    case STATUS_NORMAL:
        {
            this->DrawACell(x,y,CMaze::MAZE_SELECTED);
            this->m_nXOldFocusCell = x;
            this->m_nYOldFocusCell = y;
        }
        break;
    case STATUS_DEFINE_MAZE_START:
    case STATUS_DEFINE_MAZE_END:
        {
            if( x== 0 || x== this->m_nColumns -1 || y == 0 || y == this->m_nRows -1 )
                {// 说明当前是在围墙里面
                // 并且不是角落里
                if((y == 0 && x == 0)
                    ||(y == 0 && x == this->m_nColumns-1)
                    ||(y == this->m_nRows-1 && x == 0)
                    ||(y == this->m_nRows-1 && x == this->m_nColumns-1))
                    {// 在四个角落里，设置无效
                        this->m_nXOldFocusCell = this->m_nYOldFocusCell = -1;
```

```
                }
                else if(GetDocument()->m_pMaze->GetCellType(y,x)
                    ==CMaze::MAZE_END||GetDocument()->m_pMaze->GetCellType(y,x)
                    ==CMaze::MAZE_START)
                {
                    this->m_nXOldFocusCell = this->m_nYOldFocusCell = -1;
                }
                else
                {// 合理,则显示高亮,并设置状态
                    this->DrawACell(x,y,CMaze::MAZE_SELECTED);
                    this->m_nXOldFocusCell = x;
                    this->m_nYOldFocusCell = y;
                }
            }
            else   // 不在围墙里,设置无效
                this->m_nXOldFocusCell = this->m_nYOldFocusCell = -1;


        }
        break;
    case STATUS_DEFINE_MAZE_STONE:
        mazeRect.DeflateRect(this->m_nCellSize,m_nCellSize);
        if(mazeRect.PtInRect(point))
        {
            this->DrawACell(x,y,CMaze::MAZE_SELECTED);
            this->m_nXOldFocusCell = x;
            this->m_nYOldFocusCell = y;
            if(this->m_bMouseDown)
                this->OnLButtonDown(0,mousePoint);


        }
        else
            this->m_nXOldFocusCell = this->m_nYOldFocusCell = -1;
        break;
    default:
        ASSERT(FALSE);
    }
}
else
{
    this->m_nXOldFocusCell = this->m_nYOldFocusCell = -1;
```

    }

    CScrollView::OnMouseMove(nFlags, point);
}

/**
    [作用]   文档类的序列化操作
**/
void CMazeDoc::Serialize(CArchive& ar)
{
    if (ar.IsStoring())
    {
        //行 + 列
        int nRows = m_pMaze->GetMazeRows();
        int nColumns = m_pMaze->GetMazeColumns();
        ar<<nRows <<nColumns;

        // cellSize
        POSITION pos = this->GetFirstViewPosition();
        ar<<((CMazeView*)this->GetNextView(pos))->m_nCellSize;

        // startPos, endPos;
        ar<<m_pMaze->m_ptStart.x<<m_pMaze->m_ptStart.y
            <<m_pMaze->m_ptEnd.x<<m_pMaze->m_ptEnd.y;

        // 存储迷宫的类型——4 方向还是 8 方向
        ar<<m_pMaze->GetMazeType();
        // 算法类型
        ar<<m_pMaze->GetMazePathAlgorithmType();
        // 是否演示
        ar<<m_pMaze->GetBShowProcess();

        // other STONE cell point ;
        for(int i = 1; i < nRows -1 ; i++)
        {
            for(int j = 1; j < nColumns-1;j++)
            {
                if(m_pMaze->GetCellType(i,j) == CMaze::MAZE_STONE)
                {
                    ar<<i<<j;

```cpp
                }
            }
        }
        // 保存成功
        ((CMainFrame*)AfxGetMainWnd())->SetPaneText(_T("迷宫保存成功"));
    }
    else
    {
        POSITION pos = this->GetFirstViewPosition();
        CMazeView* pView = (CMazeView*)this->GetNextView(pos);
        // 先读取个数；
        ULONGLONG nLength = ar.GetFile()->GetLength();
        ULONGLONG readCount = 0;

        // 读取行列
        int nRows = 0;
        int nColumns = 0;
        ar>>nRows ;
        ar>>nColumns;

        pView->m_pSetDlg->m_nY = pView->m_nRows = nRows;
        pView->m_pSetDlg->m_nX = pView->m_nColumns = nColumns;

        // cellSize
        ar>>pView->m_nCellSize;
        pView->m_pSetDlg->m_nCellSize = pView->m_nCellSize;

        // 将数据还原到 CMaze 里面
        delete this->m_pMaze;
        this->m_pMaze = new CMaze(pView->m_hWnd,nRows,nColumns);

        // startPos,endPos;
        int x,y;
        ar>>x>>y;
        m_pMaze->SetMazeCellType(y,x,CMaze::MAZE_START);
        ar>>x>>y;
        m_pMaze->SetMazeCellType(y,x,CMaze::MAZE_END);

        BOOL b;                 // 4 方向还是 8 方向
        ar>>b;
```

```cpp
    m_pMaze->SetMazeDirectionType(b);
    if(b)                         // 如果是 4 方向的
    {
        ((CButton*)pView->m_pSetDlg->GetDlgItem(IDC_RADIO_4DIRECTION))->SetCheck(BST_CHECKED);
        ((CButton*)pView->m_pSetDlg->GetDlgItem(IDC_RADIO_8DIRECTION))->SetCheck(BST_UNCHECKED);
    }
    else                          // 如果是 8 方向的
    {
        ((CButton*)pView->m_pSetDlg->GetDlgItem(IDC_RADIO_4DIRECTION))->SetCheck(BST_UNCHECKED);
        ((CButton*)pView->m_pSetDlg->GetDlgItem(IDC_RADIO_8DIRECTION))->SetCheck(BST_CHECKED);
    }

    int   nType= -1;              // 算法类型
    ar>>nType;
    this->m_pMaze->SetMazePathAlgorithmType(nType);
                                  // 重新设置视图
    if(nType == CMaze::MAZE_ALGORITHM_STACK)
    {
        ((CButton*)pView->m_pSetDlg->GetDlgItem(IDC_RADIO_QUEUE))->SetCheck(BST_UNCHECKED);
        ((CButton*)pView->m_pSetDlg->GetDlgItem(IDC_RADIO_STACK))->SetCheck(BST_CHECKED);
    }
    else if(nType == CMaze::MAZE_ALGORITHM_QUEUE)
    {
        ((CButton*)pView->m_pSetDlg->GetDlgItem(IDC_RADIO_QUEUE))->SetCheck(BST_CHECKED);
        ((CButton*)pView->m_pSetDlg->GetDlgItem(IDC_RADIO_STACK))->SetCheck(BST_UNCHECKED);
    }
    else
        ASSERT(FALSE);

    //  是否演示
    ar>>b;
    if(b)
```

```cpp
        {
            m_pMaze->ChangeShowProcess();
    ((CButton*)pView->m_pSetDlg->GetDlgItem(IDC_CHECK_SHOWPROCESS))->SetCheck(BST_CHECKED);
        }
        else
    ((CButton*)pView->m_pSetDlg->GetDlgItem(IDC_CHECK_SHOWPROCESS))->SetCheck(BST_UNCHECKED);

        //  更新 pView 和  setDlg 内容.
        pView->m_pSetDlg->UpdateData(FALSE);
        pView->m_nXEndPos = pView->m_nXStartPos + pView->m_nColumns * pView->m_nCellSize;
        pView->m_nYEndPos = pView->m_nYStartPos + pView->m_nRows * pView->m_nCellSize;
        pView->m_pSetDlg->OnReleasedcaptureSlider(NULL,NULL);// 设置迷宫的演示速度
        // other STONE cell point ;
        readCount = sizeof(int)* 10;
        int i = 0,j = 0;

        while(readCount < nLength)
        {
            ar>>i;
            ar>>j;
            m_pMaze->SetMazeCellType( i,j, CMaze::MAZE_STONE);
            readCount += sizeof(int) * 2;

        }
        pView->Invalidate();
        pView->UpdateWindow();
        // 保存成功
        ((CMainFrame*)AfxGetMainWnd())->SetPaneText(_T("迷宫读取完毕"));
    }
}
```

8.3.1.5 运行测试

系统运行示意图如下图 8.5 所示。

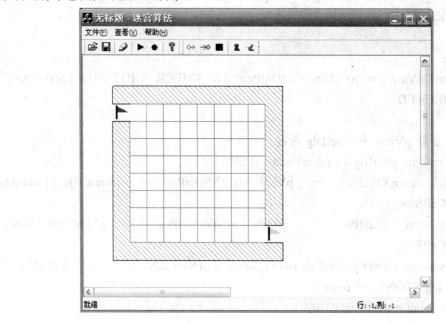

图 8.5 系统运行示意图

○系统功能菜单

系统中共有 11 项功能菜单（见图 8.5），依次为：

1）打开迷宫文件；
2）保存迷宫文件；
3）设置迷宫系统参数与数据结构；
4）运行迷宫路径搜索；
5）停止迷宫路径搜索；
6）帮助菜单；
7）设置迷宫入口位置（红旗）；
8）设置迷宫出口位置（兰旗）；
9）设置迷宫中的障碍物，再次单击则取消；
10）消除迷宫中的障碍物；
11）保存迷宫中的障碍物设置。

○设置迷宫参数

单击系统的迷宫参数设置菜单，弹出设置对话框，可对系统参数进行设置，如下图 8.6 所示。

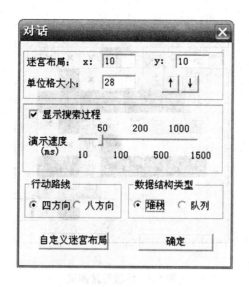

图 8.6　迷宫系统参数对话框示意图

通过迷宫参数和障碍设置，我们得到测试迷宫布局如下图 8.7 所示。

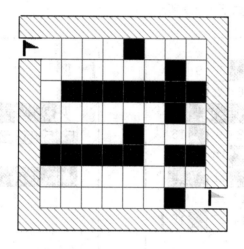

图 8.7　测试迷宫布局

堆栈算法的迷宫测试结果如下图 8.8 所示。

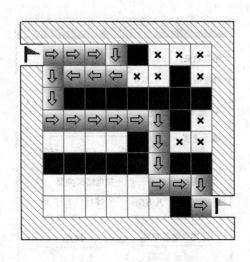

图 8.8 堆栈算法测试

在图 8.8 堆栈测试结果中显示"×"的地方表示这是一个脚印,也就是该位置在搜索时被探索过,但无法达到终点。从图中可知用该堆栈算法不一定能找到一条最快的路径。

队列算法的测试结果如下图 8.9 所示。

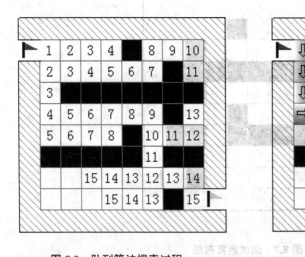 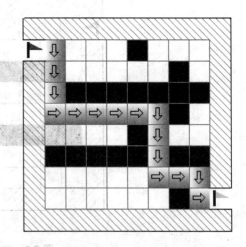

图 8.9 队列算法搜索过程　　　　图 8.10 队列算法搜索结果

图 8.9 标识了在队列搜索过程中的一些数据,右图 8.10 中将最终结果表现出来,从图中可以看出来该算法找出了一条最快的路径。

8.3.2 景区旅游信息管理系统(Console 版本)

8.3.2.1 系统简介

在旅游景区,经常会遇到游客打听从一个景点到另一个景点的最短路径和最短距离,这类游客不喜欢按照导游图的线路来游览,而是自己挑选感兴趣的景点游览。为了帮助这些游客查询信息,就需要计算出所有景点之间最短路径和最短距离。算法采用迪杰斯特拉算法或弗洛伊德算法均可。建立一个景区旅游信息管理系统，实现的主要功能包括制订旅游景点导游线路策略和制订景区道路铺设策略。

任务中景点分布是一个无向带权连通图，图中边的权值是景点之间的距离。

（1）景区旅游信息管理系统中制订旅游景点导游线路策略，首先通过遍历景点，给出一个入口景点，建立一个导游线路图。

（2）在导游线路图中，还为一些不愿按线路走的游客提供信息服务，比如从一个景点到另一个景点的最短路径和最短距离。在本线路图中将输出任意景点间的最短路径。

（3）在景区建设中，道路建设是其中一个重要内容。道路建设首先要保证能连通所有景点，花费的代价最小，这个问题可以通过求最小生成树来解决。本任务中假设修建道路的代价只与它的里程相关。

因此归纳起来，本任务有如下功能模块：

（1）创建景区景点分布图；
（2）绘制景点图输出导游线路图；
（3）求两个景点间的最短路径；
（4）输出道路修建规划图。

8.3.2.2 数据结构

```
/************************************************************************
    结点类
*************************************************************************/
class Node
{
    friend class Linklist;
    friend class Station;
    friend void Top_sort(Station &T, Linklist &L);
    friend void operate (Station &L);
public:
    Node()
    {
        s_name = '?';
        name = '?';
        leg = 0;
        flag = 0;
        next = NULL;
    }

    string s_name;                                      //出发结点名称
    string name;                                        //可到达接点名称
    float leg;                                          //可到达结点路径长度
    bool flag;                                          //访问标志位
```

```cpp
        Node *next;                                          //结点指针
};

/**************************************************************************
        链表类
**************************************************************************/
class Linklist
{
        friend class Station;
        friend void Top_sort(Station &T, Linklist &L);
        friend void operate (Station &L);
public:
        void Print_dijkstra(Linklist &L, string start, string end);     //输出最短路径
        float See_leg(Linklist &L, string name);                        //查看结点权值
        string Min_leg(Linklist &L);                                    //查找权最小的结点
        static void Change_leg(Linklist &L, float leg, string start, string end); //改变结点的权值
        void Print_Top(Node *N);                                        //输出拓扑排序结果
        void Print_Prim(Node *N);                                       //输出普利姆最小生成树
        static void Insert(Linklist *p, string name, float path);       //在两景点之间插入路径
        static int Stat(Linklist &L);                                   //统计某结点可到达结点的个数
        static bool IsT(Linklist *p, string name);                      //判断两景点之间是否存在路径
        Linklist()
        {
                s_name = '?';
                start_name = NULL;
                next_link = NULL;
        }
        string s_name;                                                  //本结点名称
        Node *start_name;                                               //可到达结点指针
        Linklist *next_link;                                            //链表指针
protected:
private:
};

/**************************************************************************
        景点类
**************************************************************************/
class Station
{
        friend class Linklist;
```

```cpp
        friend void Top_sort(Station &T, Linklist &L);
        friend void operate (Station &L);
public:
        void Change_leg(Station &L, string name);
        void Copy(Station L, Station &temp);
        void Dijkstra(Station &L, Linklist &Edge, string name);      //计算最短路径
        void Deep_visit(Station &L, Station &run, string name);      //递归遍历
        void Change_flag(Station &L, string name);                    //改变访问标志
        void Ransack(Station &temp, Station &run);                    //深度优先遍历
        static string IsHear(Station &L);              //判断是否存在入度为零的结点，并返回
        static void Delete(Station &Temp, string name);               //删除有向图中的结点
        Node * Min_path(Station L, Linklist &edge, string name);     //查找当前结点的最短路径
        void Print(Station &L);                                       //输出所有景点点和路径
        void Insert_path(Station &L);                                 //在结点之间插入路径
        void Creat_station(Station &L);                               //初始化结点
        bool IsT(Station &L, string name);                            //判断两结点之间是否存在路径
        void Min_Tree(Station L, Linklist &edge, string name);       //计算最小生成树
        int Stat(Station &L);                                         //统计结点个数
        Station()
        {
            head = NULL;
        }
        Linklist *head;                                               //图的对外指针
protected:
private:
};
```

8.3.2.3 程序清单

```cpp
/***************************************************************************
    在两结点之间插入路径
***************************************************************************/
void Linklist::Insert(Linklist *p, string name, float path)
{
    if(!Linklist::IsT(p, name)) {
        Node *N = new Node;
        N->name = name;
        N->leg = path;
        N->next = p->start_name;
        p->start_name = N;
    }
```

```cpp
        else {
            cout << "两结点之间已存在路径，无法插入新的路径！" << '\n';
        }
}

/***************************************************************************
    判断结点之间是否存在路径
****************************************************************************/
bool Linklist::IsT(Linklist *p, string name)
{
    Node *N = p->start_name;
    while(N){
        if(N->name == name)
            return true;
        N = N->next;
    }
    delete N;
    return false;
}


/***************************************************************************
    输出普利姆最小生成树
****************************************************************************/
void Linklist::Print_Prim(Node *N)
{
    if (N->next) {
        Linklist::Print_Prim(N->next);
        cout << N->s_name << "-(" << N->leg << ")->" << N->name << '\t';
    }
}


/***************************************************************************
    统计某结点可到达其他结点的个数
****************************************************************************/
int Linklist::Stat(Linklist &L)
{
    int i = 0;
    Node *N = L.start_name;
    while (N) {
        N = N->next;
```

```
            i ++;
        }
        return i;
}

/************************************************************************
    输出拓扑排序结果
************************************************************************/
void Linklist::Print_Top(Node *N)
{
    if (N->next) {
        Linklist::Print_Top(N->next);
        cout << "-->";
    }
    cout << N->name;
}


/************************************************************************
    改变结点的权值
************************************************************************/
void Linklist::Change_leg(Linklist &L, float leg, string start, string end)
{
    Node * N = L.start_name;
    while (N) {
        if (N->name == end) {
            N->s_name = start;
            N->leg = leg;
            break;
        }
        N = N->next;
    }
}

/************************************************************************
    查找权最小的结点
************************************************************************/
string Linklist::Min_leg(Linklist &L)
{
    Node * N = L.start_name;
    Node * s = new Node;
```

```cpp
            s->name = "?";
            s->leg = MAX;
            while (N) {
                if (N->flag) {
                    N = N->next;
                    continue;
                }
                if (s->leg > N->leg) {
                    s = N;
                }
                N = N->next;
            }
            return s->name;
}

/************************************************************************
        查看权值
************************************************************************/
float Linklist::See_leg(Linklist &L, string name)
{
    Node * N = L.start_name;
    while (N && N->name != name) {
        N = N->next;
    }
    return N->leg;
}


/************************************************************************
        输出最短路径
************************************************************************/
void Linklist::Print_dijkstra(Linklist &L, string start, string end)
{
    Node * N = L.start_name;
    while (N && start != end) {
        if (N->name == end) {
            Linklist::Print_dijkstra(L, start, N->s_name);
            cout << N->s_name << "-->" << N->name << '\n';
            break;
        }
        N = N->next;
```

        }
    }

/**
    更改访问标志
**/
void Station::Change_flag(Station &L, string name)
{
    Linklist *p = L.head;
    while (p) {
        Node *N = p->start_name;
        while (N) {
            if (N->name == name) {
                N->flag = 1;
                break;
            }
            N = N->next;
        }
        p = p->next_link;
    }
}

/**
    改变权值
**/
void Station::Change_leg(Station &L, string name)
{
    Linklist *p = L.head;
    while (p) {
        Node *N = p->start_name;
        while (N) {
            if (N->name == name) {
                N->leg = MAX;
                break;
            }
            N = N->next;
        }
        p = p->next_link;
    }
}

/**
    站点复制
***/
void Station::Copy(Station L, Station &temp)
{
    Linklist *p = L.head;
    while (p) {
        Linklist *q = new Linklist;
        q->s_name = p->s_name;
        q->next_link = temp.head;
        temp.head = q;

        Node *m = p->start_name;
        while (m) {
            Node *n = new Node;

            n->flag = m->flag;
            n->leg = m->leg;
            n->name = m->name;
            n->s_name = m->s_name;

            n->next = q->start_name;
            q->start_name = n;

            m = m->next;
        }
        p = p->next_link;
    }
}

/**
    初始化结点
***/
void Station::Creat_station(Station &L)
{
    cout << "请输入景点的个数：";
    int n;
    cin >> n;
    cout << "请依次输入各景点的名称：" << '\n';

```cpp
        Linklist *s = new Linklist;
        for(int i = 1; i <= n; i ++)
        {
            Linklist *p = new Linklist;
            string name;
            cin >> name;
            p->s_name = name;
            p->next_link = NULL;
            if (i == 1) {
                L.head = p;
                s = L.head;
            }
            else {
                s->next_link = p;
                s = p;
            }
        }

}

/***************************************************************************
    递归进行深度遍历
***************************************************************************/
void Station::Deep_visit(Station &L, Station &run, string name)
{
    Linklist * p = L.head;
    while (p && p->s_name != name) {
        p = p->next_link;
    }

    Node * N = p->start_name;
    while (N) {
        if (N->flag) {
            N = N->next;
            continue;
        }

        Station::Change_flag(L, N->name);
        Linklist * t = run.head;
        while (t && t->s_name != name) {
```

```
            t = t->next_link;
        }
        Linklist::Insert(t, N->name, N->leg);
        Station::Deep_visit(L, run, N->name);

        N = N->next;
    }
}

/**************************************************************************
    在有向图中删除结点
**************************************************************************/
void Station::Delete(Station &Temp, string name)
{
    Linklist *N = Temp.head;
    if (Temp.head->s_name == name) {
        Temp.head = N->next_link;
        delete N;
    }
    else {
        Linklist *s = N;
        while (N) {
            if (N->s_name == name) {
                s->next_link = N->next_link;
                delete N;
                break;
            }
            s = N;
            N = N->next_link;
        }
    }
}

/**************************************************************************
    导航最短路径函数
**************************************************************************/
void Station::Dijkstra(Station &L, Linklist &Edge, string name)
{
    Edge.s_name = name;
    Linklist * p = L.head;
```

242

```
while (p) {
    if (p->s_name == name) {
        p = p->next_link;
        continue;
    }
    Linklist::Insert(&Edge, p->s_name, MAX);
    p = p->next_link;
}

p = L.head;
while (p && p->s_name != name) {
    p = p->next_link;
}

Node * N = p->start_name;
while (N) {
    Linklist::Change_leg(Edge, N->leg, p->s_name, N->name);
    N = N->next;
}

while (Edge.Min_leg(Edge) != "?") {
    Linklist * s = L.head;
    while (s->s_name != Edge.Min_leg(Edge)) {
        s = s->next_link;
    }

    Node * N = Edge.start_name;
    while (N) {
        if (N->flag) {
            N = N->next;
            continue;
        }
        if (N->name == s->s_name) {
            N->flag = 1;
            break;
        }
        N = N->next;
    }

    Node * temp = s->start_name;
```

```cpp
        while (temp) {
            if (temp->name == Edge.s_name) {
                temp = temp->next;
                continue;
            }
            float i = s->See_leg(*s, temp->name) + Edge.See_leg(Edge, s->s_name);
            if (i < Edge.See_leg(Edge, temp->name)) {
                Edge.Change_leg(Edge, i, s->s_name, temp->name);
            }
            temp = temp->next;
        }
    }
}

/***************************************************************************
    在两结点之间插入路径
***************************************************************************/
void Station::Insert_path(Station &L)
{
    cout << "请输入要添加的两个景点及其距离：";
    string name1, name2;
    float path;
    cin >> name1 >> name2 >> path;
    if (name1 != "?" && name2 != "?" && path!=0)
    {
        if (Station::IsT(L, name1) && Station::IsT(L, name2)) {
            Linklist *p = L.head;
            while (p) {
                if (p->s_name == name1)
                    Linklist::Insert(p, name2, path);

                else if (p->s_name == name2)
                    Linklist::Insert(p, name1, path);

                p = p->next_link;
            }
            delete p;
        }
        else {
            cout << "错误的景点，请重新输入！" << '\n';
```

```cpp
        }
        Station::Insert_path(L);
    }
}

/************************************************************************
    判断该图中是否存在入度为零的结点,并返回
*************************************************************************/
string Station::IsHear(Station &L)
{
    Linklist *p = L.head;
    while (p) {
        int i = 0;
        Linklist *temp = L.head;
        while (temp) {
            if (Linklist::IsT(temp, p->s_name)) {
                i = 1;
                break;
            }
            temp = temp->next_link;
        }
        if (!i) {
            return p->s_name;
        }
        p = p->next_link;
    }
    return "false";
}

/************************************************************************
    判断是否存在此结点
*************************************************************************/
bool Station::IsT(Station &L, string name)
{
    Linklist *p = L.head;
    while (p) {
        if (p->s_name == name) {
            return true;
        }
        p = p->next_link;
```

```
        }
        delete p;
        return false;
}


/***************************************************************************
     查找当前结点的最短路径
***************************************************************************/
Node * Station::Min_path(Station L, Linklist &edge, string name)
{
        Linklist *p = L.head;
        while (p->s_name != name) {
                p = p->next_link;
        }
        Node *N = p->start_name;
        Node *s = N;
        while (edge.IsT(&edge, N->name) && N->next) {
                N = N->next;
        }
        if (N) {
                s = N;
                while (N) {
                        if (s->leg > N->leg && !N->flag) {
                                s = N;
                        }
                        N = N->next;
                }
        }
        return s;
}


/***************************************************************************
     计算普利姆最小生成树
***************************************************************************/
void Station::Min_Tree(Station L, Linklist &edge, string name)
{
        Linklist *p = L.head;
        while (p->s_name != name) {
                p = p->next_link;
        }
```

```cpp
        Node *start = new Node;
        start->name = name;
        start->s_name = name;
        start->leg = 0;
        start->next = edge.start_name;
        edge.start_name = start;
        L.Change_flag(L, name);
        L.Change_leg(L, name);

        while (edge.Stat(edge) < L.Stat(L)) {
            Node *t = edge.start_name;
            Node *s = t;
            while (t) {
                if ((Station::Min_path(L, edge, s->name)->leg > \
                    Station::Min_path(L, edge, t->name)->leg) && \
                    !Station::Min_path(L, edge, t->name)->flag) {
                    s = t;
                }
                t = t->next;
            }
            Node *N = new Node;
            N->s_name = s->name;
            N->name = Station::Min_path(L, edge, s->name)->name;
            N->leg = Station::Min_path(L, edge, s->name)->leg;
            N->next = edge.start_name;
            edge.start_name = N;
            L.Change_flag(L, N->name);
            L.Change_leg(L, N->name);
        }
    }

/*****************************************************************************
    输出所有景点和路径
*****************************************************************************/
void Station::Print(Station &L)
{
    Linklist *p = L.head;
    while (p) {
        Node *N = p->start_name;
```

```cpp
        cout << p->s_name;
        while (N) {
            cout << "-->" << N->name << '(' << N->leg << ')';
            N = N->next;
        }
        cout << '\n';
        delete N;
        p = p->next_link;
    }
    delete p;
}

/************************************************************************
    深度优先遍历
*************************************************************************/
void Station::Ransack(Station &temp, Station &run)
{
    Linklist * p = run.head;
    while (p) {
        p->start_name = NULL;
        p = p->next_link;
    }
    Station::Change_flag(temp, temp.head->s_name);
    Station::Deep_visit(temp, run, temp.head->s_name);
}

/************************************************************************
    统计结点个数
*************************************************************************/
int Station::Stat(Station &L)
{
    int i = 0;
    Linklist *p = L.head;
    while (p) {
        i ++;
        p = p->next_link;
    }
    delete p;
    return i;
}
```

/***
    进行拓扑排序
***/
```cpp
void Top_sort(Station &T, Linklist &L)
{
    string name;
    while (T.head) {
        if (Station::IsHear(T) != "false") {
            name = Station::IsHear(T);
        }
        else{
            cout << "该图中存在有向环！" << '\n';
            return;
        }
        L.Insert(&L, name, 0);
        Station::Delete(T, name);
    }
    cout << "景区导航路线拓扑图为：\n";
    L.Print_Top(L.start_name);
}
```

/***
    主函数
***/
```cpp
void main ()
{
    cout<<"\n**            景区旅游信息管理系统              **"<<endl;
    cout<<"=============================================="<<endl;
    cout<<"**            A --- 构建景区图                  **"<<endl;
    cout<<"**            B --- 导航线路图                  **"<<endl;
    cout<<"**            C --- 导航最短路径                **"<<endl;
    cout<<"**            D --- 道路修建图                  **"<<endl;
    cout<<"**            E --- 退出   程序                 **"<<endl;
    cout<<"=============================================="<<endl;
    Station L;
    L.Creat_station(L);
    L.Insert_path(L);
    char ch = ' ';
    while( ch!='E' ){
```

```cpp
cout<< '\n' <<"请选择操作 : ";
cin >> ch;
switch(ch) {
case 'A':
    {
        cout << "各景点到其他景点的景区旅游图为: " << endl;
        Station temp;
        temp.Copy(L, temp);
        temp.Print(temp);
        break;
    }
case 'B':
    {
        Station temp, run;
        temp.Copy(L, temp);
        run.Copy(L, run);

        run.Ransack(temp, run);
        run.Print(run);

        Linklist Top;
        Top_sort(run, Top);
        cout << '\n';
        break;
    }
case 'C':
    {
        Station temp;
        temp.Copy(L, temp);

        Linklist min_path;
        cout << "请输入现在所在景区以及目的景区: ";
        string start, end;
        cin >> start >> end;
        temp.Dijkstra(temp, min_path, start);
        cout << "最短的路线是: " << '\n';
        min_path.Print_dijkstra(min_path, start, end);
        Node * N = min_path.start_name;
        while (N && N->name != end) {
            N = N->next;
```

```
                }
                cout << "最短的距离是：" << N->leg << '\n';
                break;
        }
    case 'D':
        {
            Station temp;
            temp.Copy(L, temp);

            Linklist Prim;
            temp.Min_Tree(temp, Prim, L.head->s_name);
            cout << "景区线路修建规划最佳方案为：" << '\n';
            Prim.Print_Prim(Prim.start_name);
            cout << '\n';
            break;
        }
    case 'E':
        break;
    default:
        {
            cout << "输入错误，请选择正确的操作！" << '\n';
            break;
        }
    }

}
}
```

8.3.2.4 运行测试

```
**          景区旅游信息管理系统          **
==========================================
**           A --- 构建景区图            **
**           B --- 导航线路图            **
**           C --- 导航最短路径          **
**           D --- 道路修建图            **
**           E --- 退出  程序            **
==========================================
请输入景点的个数：4
请依次输入各景点的名称：
a b c d
请输入要添加的两个景点及其距离：a b 6
请输入要添加的两个景点及其距离：a c 7
请输入要添加的两个景点及其距离：a d 9
请输入要添加的两个景点及其距离：b c 4
请输入要添加的两个景点及其距离：c d 1
请输入要添加的两个景点及其距离：b d 3
请输入要添加的两个景点及其距离：? ? 0

请选择操作：A
各景点到其他景点的景区旅游图为：
d-->a<9>-->c<1>-->b<3>
c-->a<7>-->b<4>-->d<1>
b-->a<6>-->c<4>-->d<3>
a-->b<6>-->c<7>-->d<9>

请选择操作：B
d-->a<9>
c
b-->c<4>
a-->b<6>
景区导航路线拓扑图为：
d-->a-->b-->c

请选择操作：C
请输入现在所在景区以及目的景区：a d
最短的路线是：
a-->c
c-->d
最短的距离是：8

请选择操作：D
景区线路修建规划最佳方案为：
a-<6>->b        b-<3>->d        d-<1>->c
```

8.3.3 景区旅游信息管理系统（MFC 版本）

8.3.3.1 系统简介

在旅游景区，经常会遇到游客打听从一个景点到另一个景点的最短路径或最短距离，这类游客不喜欢按照导游图的线路来游览，而是挑选自己感兴趣的景点游览。为了帮助这些游客查询信息，就需要计算出所有景点之间最短路径和最短距离。算法采用迪杰斯特拉算法或弗洛伊德算法均可。建立一个景区旅游信息管理系统，实现的主要功能包括制订旅游景点导游线路策略和制订景区道路铺设策略。

任务中景点分布是一个无向带权连通图，图中边的权值是景点之间的距离。

（1）景区旅游信息管理系统中制订旅游景点导游线路策略，首先通过遍历景点，给出一个入口景点，建立一个导游线路图。

（2）在导游线路图中，还为一些不愿按线路走的游客提供信息服务，比如从一个景点到另一个景点的最短路径和最短距离。在本线路图中将输出任意景点间的最短路径。

（3）在景区建设中，道路建设是其中一个重要内容。道路建设首先要保证能连通所有景点，但又要付出的代价最小，可以通过求最小生成树来解决这个问题。本任务中假设修建道路的代价只与它的里程相关。

因此归纳起来，本任务有如下功能模块：

（1）创建景区景点分布图；
（2）绘制景点图输出导游线路图；
（3）求两个景点间的最短路径；
（4）输出道路修建规划图。

8.3.3.2 设计流程

（1）界面设计

程序中的窗口框架设计如下图 8.11 所示。

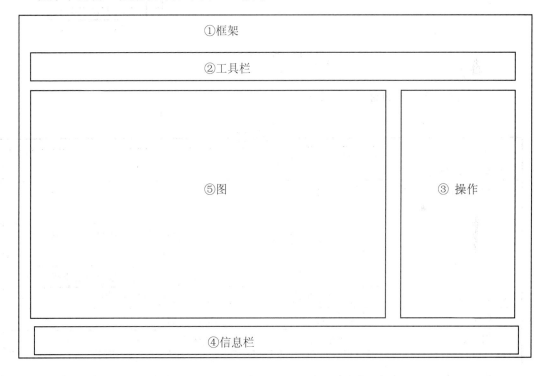

图 8.11　窗口框架设计图

整个程序采用 MFC 的单文档模板，基本框架类采用默认类名——CMyApp，CMyDocument，CFrameWnd，CMyView。然后以 CMyView 作为一个基本容器，在它里面添加子窗口。

所有的窗口类关系如下图 8.12 所示。

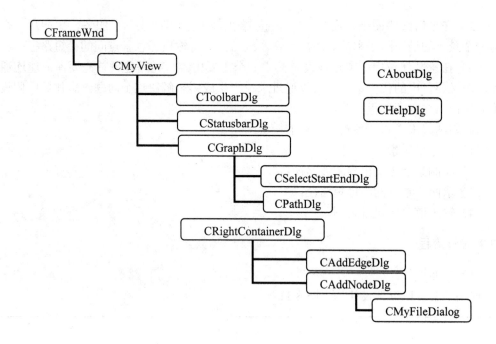

图 8.12 窗口类关系图

程序中的各个类的意义如下表：

表 8.1 类的描述表

类名 父类	类 型	意 义
ALGraph	数据结构封装	邻接表
EdgeNode	数据结构算法辅助结构	封装一个结点和与该结点的一条边。被用于邻接表中，最小生成树中，等。
Tree	数据结构封装	生成向导图中，用于穷举的生成树。
CGraph	数据结构封装	封装了所有图的算法及操作
CNode	数据结构封装	封装图中顶点
CEdge	数据结构封装	封装图中的边
窗口类		
CAboutDlg	对话框—弹出窗口	关于对话框
CAddEdgeDlg	对话框—子窗口	提供对图中边的添加、修改、删除的功能。
CAddNodeDlg	对话框—子窗口	提供对图中顶点的添加、修改、删除的功能。
CGraphDlg	对话框—子窗口	1.接收从 CAddEdgeDlg 传递过来的操作。 2.接收从 CAddNodeDlg 传递过来的操作。 3.负责 ini 文件的读写操作。 4.负责图的界面显示和操作。 5.CGraph 的对象在这里面声明。

续表

类名 父类	类 型	意 义
CHelpDialog	对话框-弹出窗口	帮助，提供程序的操作方法。
CMyFileDialog :CFileDialog	对话框-弹出窗口	重新设计打开文件对话框，提供图片预览功能。
CPathDlg	对话框-弹出窗口	将 CGraph 中生成的向导图路线输出。
CRightContainerDlg	对话框-子窗口	原本用于将在程序窗口右边要显示的窗口（如 CAddEdgeDlg，CAddNodeDlg）都作为它的子窗口以便于管理。
CSelectStartEndDlg	对话框-弹出窗口	在于在生成向导图前选择向导图的起点和终点。
CStatusbarDlg	对话框-子窗口	用于在程序底部显示一些提示信息。
CToolbarDlg	对话框-子窗口	用于在程序顶部提供一些控制按钮。
框架类		
CMainFrame	框架类	实现全屏功能
CMyApp	框架类	应用类，相当于一个全局变量，在里面存放了一些共用数据作为全局变量。实现程序状态的设置。
CMyDoc	框架类	无
CMyView	框架类	用于协调各个子窗口的位置。
自绘类		
CFreshBrush :CBrush	DLL 接口文件	自定义的一个用于实现渐变画刷的一个类。
CMenuItemContext	自绘菜单辅助类	保存自绘菜单的所有有关信息
CMyBmpButton :CButton	自绘按钮类	实现按钮的自绘。 （封装的不好，待改进…）
CMyCoolMenu :CMenu	自绘菜单类	实现菜单的自绘。 （还存在一些问题…）
CNonClientMetrics	自绘菜单辅助类	
CPopupText	自绘菜单辅助类	实现自绘菜单中的提示条。
COwnerDrawListBox :CListBox	自绘列表框类	实现列表框的自绘

为了能够快速实现窗口之间信息的交互，在使用::SendMessage 的时候，程序保存了一些变量作为 CMyApp 的成员，也就相当于全局变量。如：

// 这里保存对话框的句柄，用于在程序中向其他对话框发送消息

```
HWND        m_hwndGraphDlg;
HWND        m_hwndAddNodeDlg;
HWND        m_hwndAddEdgeDlg;
HWND        m_hwndMyView;
```
例如在点击了工具栏中的"向导图"时，程序代码为：

```
extern CMyApp theApp;
::SendMessage(theApp.m_hwndGraphDlg, WM_CREATE_GUIDEMAP, 0, 0);
```

这样就直接向 CGraphDlg 发送了一条生成向导图的消息。程序中由于用对话框将各个模块进行分隔，也造成相互之间通信的麻烦。因此程序中大量使用了::SendMessage 语句进行通信。所有的消息定义都存放在 CMyApp 类的声明文件中，作为全局变量使用。

（3）程序状态

为了实现不同操作状态下的界面显示，设置一个程序状态变量作为全局变量：

```
private:
    int       m_nCurrentStatus;      // 程序当前的状态
public:
    int       GetCurrentStatus()
    {
        return m_nCurrentStatus;
    }
    void SetCurrentStatus(int    nNewStatus);
```

每次调用 theApp.SetCurrentStatus 后，也将框架标题作相应改变，标明当前所处的状态。具体的各个状态意义如下：

```
enum Status
{
    Normal      = 0,          // 一般

    Log         = 2,          // 正在登录（未实现）
    Manage      = 3,          // 已登录

    AddNode     = 4,          // 添加景点
    ModifyNode  = 5,
    DeleteNode  = 6,
    AddEdge     = 7,          // 添加路径
    ModifyEdge  = 8,
    DeleteEdge  = 9,

    GuideMap    = 10,         // 生成向导图
    FreePath    = 11,         // 自由路线
    BuildMethod = 12          // 景区建设方案
};
```

（4）图的界面设计

图的显示界面分为两块——顶点和边，分别由以下两个函数完成：

void　　CNode::DrawNode(CDC &memDC，CDC &bufDC)；

void　　CEdge::DrawEdge(CDC& memDC，CGraph& g)。

界面的显示和操作都交由 CGraphDlg 完成，然后调用 DrawNode 和 DrawEdge 显示出来。

8.3.3.3 数据结构

○图的组成

在数据结构中，图是由顶点和边组成的。在这里顶点代表景区的各个景点，边代表某两个景点之间的路径。因此这里的图是一个无向图。

○顶点

程序中顶点中数据内容及其意义如下表：

表 8.2　顶点信息描述表

数据类型	变量名	意　义
算法数据		
int	m_nodeIndex	一个顶点的索引值
int	m_nFlag	顶点数据，主要被用来在各种算法中标记该顶点是否被搜索过或者是否符合条件
基本属性		
CString	m_nodeName	顶点名称（景点名称）
CString	m_nodeIntroduce	景点介绍（未实现）
int	m_nNodeType	顶点的类型，大致分为了 8 种，最主要的几种是<入口，出口，景点，分路口，…，其他>
int	m_xRatio	顶点坐标
int	m_yRatio	这是一个千分比，比如（500，500）表示的是（500‰，500‰），也就是中央位置，这样主要是为了实现界面的伸缩性。
int	m_xBrowseRatio	景点缩略图坐标位置
int	m_yBrowseRatio	意义同上
NodeStatus	m_nCurrentNodeStatus	描述顶点当前的状态。状态包括（正常，左键选中，右键选中）

续表

数据类型	变量名	意　义
资源数据		
BrowseBmpType	m_nBrowseBmpType	缩略图的类型
HBITMAP	m_hbBrowseBmp	缩略图句柄
BITMAP	m_bmBrowseBmp	缩略图属性
CArray<HBITMAP,HBITMAP&>	m_hbNodePictures	景点通览图列表（未实现）
CArray<BITMAP,BITMAP&>	m_bmNodePictures	景点通览图属性列表（未实现）
其他		
CRectTracker	m_rtNode	用于顶点的选取和移动
CRectTracker	m_rtNodeBrowse	同上

○边

程序中边的数据内容及其意义如下：

表 8.3　边信息描述表

数据类型	变量名	意　义
算法数据		
int	m_edgeIndex	一条边的索引值
int	m_nFlag	用与在算法过程中，设置与边相关的数据。主要用在遍历过程中标识是否已经被搜索过或者是否符合要求。
基本属性		
int	m_nodeIndex1	该边的一个顶点的索引
int	m_nodeIndex2	该边的另一个顶点的索引
double	m_pathLength	路径长度（权）
EdgeStatus	m_nCurrentEdgeStatus	边的状态（是否被选中）
int	m_pathWidth	路径的宽度
bool	m_bPathIsLine	（未实现）
资源数据		
COLORREF	m_pathColRGB	路径的颜色
bool	m_bPathIsLine	是否存在路径

在程序运行过程中，所有的顶点和边的数据都被保存在类 CGraph 中。

```
CArray<CNode,CNode&>    m_nodeArray;
CArray<CEdge,CEdge&>    m_edgeArray;
```
这里采用 MFC 提供的 CArray 数组类保存这些数据，然后在任何其他地方要引用这些数据的时候，都是采用引用地址的方法，以免又生成一个新的对象。例如：
```
for(int j = 0; j < size; j ++)
{
    for(int k = 0; k < size; k ++)
    {
        CNode* pNode1 = &this->m_nodeArray[j];
        CNode* pNode2 = &this->m_nodeArray[k];
        AdjMatrix[j][k] =
            this->GetOneEdgeBy2Nodes(pNode1,pNode2);
    }
}
```
这是生成邻接矩阵的一段代码，矩阵的二维数组中存储的是两个顶点之间的边的地址。为了获取顶点和边的数据，从 CArray 中引出地址：&m_nodeArray[j]，GetOneEdgeBy2Nodes 返回的数据也是一个 CEdge 的指针。也就是说整个程序运行过程中只有一份数据全部放在 CGraph::CArray<…>中，其他地方的数据全都是引用数组中的地址。

〇邻接表
```
/**************************************************************************
    邻接表（ALGraph）
**************************************************************************/
```
邻接表是由 n 条（n 等于图中顶点个数）链表组成的一个存储方式，这 n 条链表又是一个数组。链表中的每个结点存储的数据为：
```
// 描述邻接表的结点信息
struct   EdgeNode
{
    CNode*    pNode;        // 该结点
    CEdge*    pEdge;        // 为 NULL 或者 边
};
```
整个邻接表可表示为：
```
// 由邻接链表组成的图
class ALGraph
{
public:
    typedef CList<EdgeNode,EdgeNode&> AdList;
    AdList* pList;              // 链表数组

    int     nNodeCount;         // 图中结点个数
    ... ...
```

};
○邻接表示意图

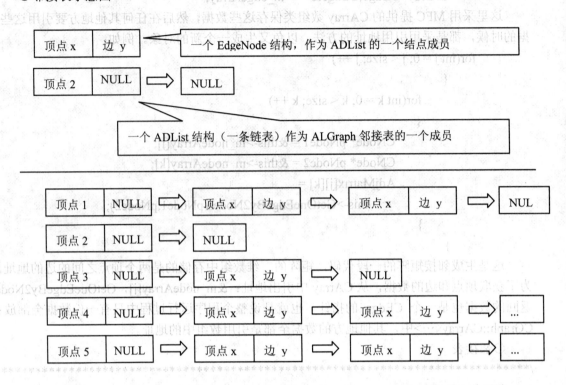

图 8.13 图的邻接链表示意图

从图 8.13 中可以看出，每个链表的表头，代表着一个顶点，它后面的顶点 x、边 y 表示的是顶点 n 与顶点 x 之间的关联——边 y。边 y 中封装了两个顶点之间的距离（权）。

○创建邻接表

创建邻接表依赖于在 CGraph:: m_nodeArray 和 CGraph:: m_edgeArray 中保存的所有图的数据，ALGraph 中仅引用其中的地址。创建邻接表的函数为：

void CGraph::CreateAdjacencyListGraph（ALGraph& g）。

其大致流程为：

（1）分配 n 个顶点的链表空间；

（2）遍历 CGraph::m_nodeArray，初始化各个链表的表头；

（3）遍历 CGraph::M_edgeArray，将每条边的顶点数据分别添入到相应的链表当中。

○邻接矩阵

/**

邻接矩阵（CEdge***   AdjMatrix）

**/

图的邻接矩阵就是一个二维数组，用各个顶点作为索引，用两个顶点之间的边作为数组里的内容。CEdge***   AdjMatrix 表示 AdjMartrix 是一个二维数组，数组里面的内容为 CEdge*（引用 CGraph::m_edgeArray 中的地址）。

在这里需要注意的是 AdjMartrix[x][y]，这里的 x、y 是 CGraph::m_nodeArray 这个数组中的索引。比如 AdjMatrix[2][3]代表的是 m_nodeArray[2]这个顶点与 m_nodeArray[3]这个顶点之间的边，由两个顶点获取他们相关联的边调用：

CEdge* CGraph::GetOneEdgeBy2Nodes(CNode* sideNode1，CNode* sideNode2)

如果两个顶点之间有边相连，那么可以认为 AdjMartrix[2][3]代表一个 CEdge*，否则 AdjMartrix[2][3] = NULL。

○邻接矩阵的创建过程分为两步
（1）初始化 AdjMatrix 二维数组；
（2）往数组里面填充内容。
函数代码为：

```
void CGraph::CreateAdjacencyMatrix(CEdge*** &AdjMatrix)
{
    typedef CEdge*    AMItem;
    int size = this->m_nodeArray.GetSize();
    // 构造
    AdjMatrix = new AMItem*[size];
    for(int i = 0; i < size ; i++)
    {
        AdjMatrix[i] = new AMItem[size];
    }
    // 初始化
    for(int j = 0; j < size; j ++)
    {
        for(int k = 0; k < size; k ++)
        {
            CNode* pNode1 = &this->m_nodeArray[j];
            CNode* pNode2 = &this->m_nodeArray[k];
            AdjMatrix[j][k] =
                this->GetOneEdgeBy2Nodes(pNode1,pNode2);
        }
    }
}
```

○关于数据的序列化
为了实现数据的存储，程序中采用了传统的 ini 文件方式（相当于一个独立的注册表）。
我们可将整个系统的数据分成两部分，一个为系统数据，用来设置这个程序的一些属性；另一个为图的数据，包括图的顶点、图的边。

○系统数据
系统数据文件名为 setting.ini，里面的数据存储格式如表 8.4 所示。

表 8.4 系统数据数据存储表

section	key	value_解释
景区	name	景区名称
	remind	在滚动栏中要显示的信息.
皮肤	skinType	程序要使用的皮肤类型（未实现） value 值为工程目录下的 skin 文件夹里皮肤文件夹的名称.
图文件	path	图数据文件的文件名. 要求图数据文件（ini）也在同一目录中.
	nextNodeIndex	顶点索引的计数器 每添加一个顶点，都从这里读取一个索引值，以保证每个索引的唯一性.
	nextEdgeIndex	边索引的计数器 (同上)

上表 8.4 "图文件"的"path"用来设置图数据文件的文件名，这样作是为了实现能够将图数据独立出来，单独作为一个模块。因此只需要通过设置 path 的值，就能在不同的图数据文件中进行切换。

○图数据

这里存储着所有的顶点和边的信息。顶点的 section 命名规则为：

    node + 空格 + 顶点索引

边的 section 命名规则为：

    edge + 空格 + 边索引

○关于程序中文件的引用

程序中对于路径的引用全部采用绝对路径的方式，以免采用相对路径而产生错误。

系统数据 ini 文件被与 CMyApp 进行了关联，这样能直接调用 CWinApp 的一些函数来读写文件。

free((void*)theApp.m_pszProfileName);

theApp.m_pszProfileName = _tcsdup(theApp.m_strProjectDirectory + _T("\\setting.ini"));

// 禁止调用注册表的功能，改用外部 ini 文件存取

theApp.m_pszRegistryKey=NULL;

图文件数据的读写采用一系列 API 函数实现。如 GetPrivateProfileString，GetPrivateProfileInt，WritePrivateProfileString，GetPrivateProfileSectionNames。

8.3.3.4 程序核心算法说明

/***

  图的连通性

***/

在进行其他图的操作之前，必须首先判断是否满足——图是一个连通图，从图中的任一个顶点出发，能够到达其他的任意一顶点。

判断图的连通性采用的方法是：广度遍历该图，经过一次遍历后，判断所有的顶点是否都被遍历到，如果存在没被遍历到的顶点则说明该图不是连通图。因为遍历中要求知道从一个顶点出发能够到达的其他顶点，所以这里的图的存储方式采取邻接表。

判断图的连通性的函数为：

bool CGraph::IsConnectedGraph(ALGraph& g)

/***
向导图的生成
***/

○算法思路

向导图的一些准则：

在景区，应该至少有一个入口处和若干个出口处。

当从 A->B 以后，而 B 不能再通往别处时，应该再从 B 返回到 A，这时 B->A 也要算在路程当中。

从景区出来的时候，也允许从入口处出来。

导游图要作到从入口出发，从出口出来，并能达到路程最小。

○关于为什么不能用深度搜索来生成导游图

深度搜索不能解决及时地找到出口。

由于景区图是一个无向连通图，虽然深度搜索最终能回到入口点，但是在搜索过程中出现出口时不能很好地作出处理。

深度搜索不能解决回溯时也算一条路。

深度搜索不能保证一定是最快的路。

○怎样用树来解决向导图的生成

树结点中的内容为：

```
struct Tree
{
    CNode*      pNode;          // 当前结点
    CEdge*      pEdge;          // 子节点与父节点之间的边
    Tree*       pParent;        // 父结点
    double      nPathLength;    // 到目前为止，从根结点到当前结点的路径长
};
```

用一个队列来协助树的生成，队列里的元素称为活动结点。

deque<Tree*> activeNodeDeque;   // 队列用来记录还未扩展的结点

活动结点是树中那些还没有生成它的子结点的树结点，每次生成的子结点都将被加入到活动结点队列，并将父结点（就是当前队列中的第一个元素）从活动结点队列中出队。当一个活动结点从队列中出队，要不就是因为已被扩展，变成了父结点，要么就是因为被截枝，不符合条件（同时这个结点所在的路线不再被考虑）。当找到一个结果时，将该结点保存。由该结点往上搜索，就能得出向导路线。

树的构造原则是每次从活动结点中取出队首元素 Tree* parentNode，将 parentNode->pNode 这个结点在图中的所有相邻结点作为生成树中 parentNode 的子结点。

当一个子结点满足——

①子结点是出口；
②子结点所在的树分支路线中，走过了所有的景点。
这个时候该条分支路线就算一条向导路线，但还不一定是最佳路线。在每一个树结点中保存着从根结点到当前结点的路线总长，通过比较长度，就能对各条路线进行合理地选取。
○算法的有效性分析
该算法用的是穷举法，将所有可能的路线在树中表示出来，随着算法的进行，树一直在扩大，因此树的那些叶子结点的 Tree::nPathLength 值也在增大。因为这些路线中一定会有一条是最佳的向导路线，也就是有一个确定的最短的路线长度 bestLength，用这个 bestLength 来作树的截枝条件，将所有路线长度大于 bestLength 的结点出队。整个算法的循环是依据活动结点的队列是否为空来进行的，当活动结点为空后，循环退出。
○算法描述
罗列出景区的所有入口及出口，选出一个入口和一个出口。其中出口的选择允许出口也是一个入口，或者不确定一个固定的出口，由程序算法自动求出一个最快的出口。
①创建邻接表。
②判断图的连通性。
③开始以树的形式生成解决方案。以入口点作为整棵树的根 ROOT，并加入活动结点队列。
④取出队列中的头一个树结点，开始扩展它。
从邻接表中获取到与它相邻的顶点，对获取到的顶点进行判断（裁枝）：
如果获取到的新顶点后，总路径长度大于最目前获取到的最佳路径长度，舍弃；
如果是走了一条和上次重复的路线，舍弃。例如从 A->B->A->B，最后那个 B 就算是一条重复路线。
⑤新结点符合要求后，构造出它的树结点结构 Tree*，将它的父结点设置为从队列头获取到的那个树结点，并将它作为一个活动结点入队。
⑥对获取到的新顶点进行出口判断，判断依据为：
该顶点属于在 a 中设置的一个出口；
该顶点所有的树的分支已经走过了所有的景点。
那么该条分支路径是一条向导图，但不一定是最快的路线。
⑦如果新顶点出口判断成功，进行下列操作：
如果目前还没有获取到一条向导路线，将该条路线作为当前最佳路线。
如果该条路线比最佳路线还短，那么将该条路线作为最佳路线保存起来，并记录它的长度。同时扫描活动结点中所的有树结点，如果这些树结点的长度已经大于新生成的这条最佳路线，将它们从活动结点中出队。
如果该条路线和最佳路线长度一致，那么该条路线一起保存起来作为另一条最佳路线最后一走提交给用户选择。
⑧处理完列队中头结点的所有子结点后，将头结点出队。重复④。
⑨直到活动结点队列为空后，最后保存的那条路线就是我们所在的向导图。
⑩用对话框的形式输出结果。
⑪释放所有资源。
○具体的实现参见代码及其注释

bool CGraph::CreateGuideMap(int nIndexStart, int nIndexEnd)
/**
最短单源路径
**/

○算法思路

这里实现了自由旅行向导，允许游客任意起点和终点，求出它们之间的最短路径的走法。程序中也采用迪杰斯特拉（Dijkstra）算法来实现。

○算法描述

首先图必须是一个连通图，使用 IsConnectedGraph 函数判定。

在该算法中图的存储方式采用的是邻接矩阵，因为需要知道任意两点之间的路径。

算法中需要的一些辅助变量参见表 8.5。

表 8.5 最短单源路径辅助变量描述表

变量类型	变量名	意　义
CEdge***	AdjMatrix	邻接矩阵
bool*	bSelected	因为单源最短路径算法其实也是求出了从起点到其他所有点的最短路径，算法每一次循环就能得出从起点到其中一个顶点的最短路径。bSelect 就是用来标识这些顶点是否已被求出它的最短路径。如 bSelect[0]=true 则表示从起点到顶点 m_nodeArray[0]的最短路径被求出。
double*	shortestPaths	记录各个顶点到起点最短路径（从起点到该点的路径长度），这是算法重点进行更新和计算的变量。 shortestPahts[i]=0 表示顶点 i 到起点的最快路径还未知。
int*	prevIndex	用于得出从顶点 i 到起点的最快路径所要经过的边。比如 prevIndex[0]= 1，表示顶点 0 要到起点的最快路线中，要先到达顶点 1，然后再取出 prevIndex[1]的值，假设 prevIndex[1]=2，就说明要到达起点的最快路线，顶点 0 要先到顶点 1，再到顶点 2，……，一直推导到 preIndex[x]=起点为止。

例如对于下图 8.14 中的一个无向图，该算法的过程演示如下：

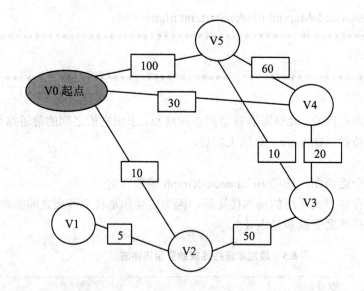

图 8.14 算法过程演示图（1）

步骤	被选景点	bSelected	shortestPaths	prevIndex
1	V0	{1,0,0,0,0,0}	{0, 0,10, 0,30,100}	{0,-1,0,-1,0,0}
2	V2	{1,0,1,0,0,0}	{0,15,10,60,30,100}	{0,2,0,2,0,0}
3	V1	{1,1,1,0,0,0}	{0,15,10,60,30,100}	{0,2,0,2,0,0}
4	V4	{1,1,1,0,1,0}	{0,15,10,50,30,90}	{0,2,0,4,0,4}
5	V3	{1,1,1,1,1,0}	{0,15,10,50,30,60}	{0,2,0,4,0,3}
6	V5	{1,1,1,1,1,1}	{0,15,10,50,30,60}	{0,2,0,4,0,3}

图 8.14 算法过程演示图（2）

在这个例子中，假如我们需要求从起点 V0 到 V3 的最短距离，需先找出 prevIndex[3] = 4，再找到 prevIndex[4]=0，即找到最快路径为 V3 <-> V4 <-> V0。

算法中的关键代码为：

```
for(int j = 0; j < size  ; j ++)
{
    startIndex = this->updateShortestPath(
        bSelected,shortestPaths,AdjMatrix,prevIndex,size,startIndex);
}
```

updateShortestPath 的作用是每次从 shortestPaths 中挑出一个最短的且 bSelect[i]==0 的结点出来，并且更新 bSelected、shortestPaths、prevIndex 的值，整个过程循环 size 次（size 为顶点个数）。

〇具体代码

参见：bool CGraph::ShortestPath_DIJ(CNode *startNode, CNode *endNode)。

/**
最小生成树
**/

○算法思路

景区建设方案其实就是应用图中的最小生成树算法来实现的，使用最小的花费，实现各个景点之间的连通。

程序中的最小生成树算法采用的是普里姆算法。见附录参考文献［1］第 173～175 页。

○算法描述

算法中的辅助变量有：

表 8.6 最小生成树辅助变量描述表

变量类型	变量名	意　义
CEdge***	AdjMatrix	邻接矩阵
vector<int>	U	已被选择的顶点的集合
EdgeNode*	closedge	closedge[i]->pNode 表示当前结点，closedge[i]->pEdge 表示：顶点 i 到 U 中的所有顶点中的最短的那条路径，如果没有路径，那么 pEdge = NULL

○算法流程

①连通性判断；

②创建邻接矩阵；

③初始化 U 和 closedge——

如果 closedge[i].pNode == NULL，表示该点已在被选中了，在最小生成树的结点当中，这个时候就会忽略 closedge[i].pEdge；如果 closedge[i].pNode != NULL，closedge[i].pEdge == NULL，表示现在还不能到达该结点。

```
void CGraph::InitClosedge(CEdge*** &AdjMatrix,EdgeNode* closedge)
{
    int size = this->m_nodeArray.GetSize();
    closedge[0].pNode = NULL;      // 默认第一个已经被选中到 U 中
    for(int i = 1; i < size; i ++)
    {
        closedge[i].pNode = &this->m_nodeArray[i];
        closedge[i].pEdge = AdjMatrix[0][i];
    }
}
```

④获取 closedge 中的最短路径

```
// 求出这次辅助结构中最小的那条边
// [参数]
//      size  结构里的元素个数
// [返回] 辅助结构中最短的那一项
```

```cpp
int    CGraph::getClosestEdge(EdgeNode* closedge,int size)
{
    double shortest = -1;          // 最短路径长
    int index = -1;                // 最短的那项
    for(int i = 0; i < size ; i ++)
    {
        if(closedge[i].pNode != NULL)
        {
            if(closedge[i].pEdge != NULL)
            {
                if(shortest == -1 ||
                    shortest > closedge[i].pEdge->m_pathLength)
                                   // 找到一项
                {
                    index = i;
                    shortest =   closedge[i].pEdge->m_pathLength;
                }
            }
        }
    }
    return index;
}
```

⑤将获取的最小的 closedge[i] 的顶点序号加入到 U 中，并更新 closedge 中的数据
```cpp
        // 将最小的那条边的结点放入 U 中
        U.push_back(index);
        TRACE1("结点<%s>被选中",closedge[index].pNode->m_nodeName);
        closedge[index].pNode = NULL;
        // 标识这条路径，用于界面显示
        closedge[index].pEdge->m_nFlag =(int)true;
        // 更新 closedge
        for(int i = 0 ; i < size ; i ++)
        {
            if(closedge[i].pNode == NULL)           // 该点已在 U 中
                continue;
            int uSize = U.size();

            for(int j = 0; j < uSize; j ++)
            {
                CEdge* pEdge = AdjMatrix[i][U[j]];
                // U[j] 为相应已被选入最小生成树结点的序号
```

```
            if(pEdge == NULL)
                continue;
            if(closedge[i].pEdge == NULL)
                closedge[i].pEdge = pEdge;
            else
            {
                if(closedge[i].pEdge->m_pathLength > pEdge->m_pathLength)
                    closedge[i].pEdge = pEdge;
            }
        }
    }
```
⑥重复步骤④，直到 U.size() == 顶点的个数；
⑦释放资源；
⑧刷新界面显示结果。
〇具体代码

参见：bool CGraph::CreateMinSpanTree()。

8.3.3.5 运行测试

（1）正常状态下的程序主界面：

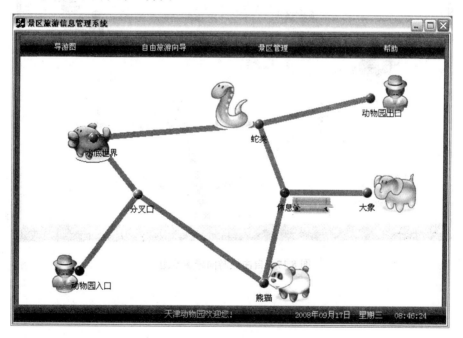

图 8.15 程序界面演示图

(2)生成向导图的界面：

图 8.16 程序界面演示图

(3)自由旅游向导图测试界面如下：

图 8.17 自由旅游向导演示图

(4) 修改景点界面操作如下图 8.18 所示。

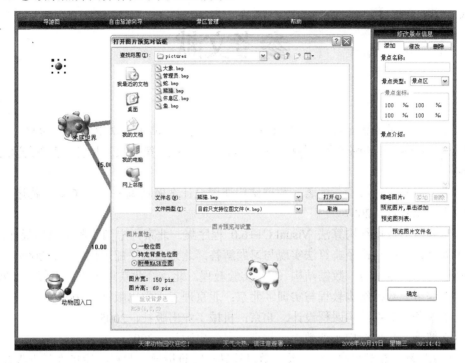

图 8.18 景区建设方案图

(5) 景区建设方案图测试界面如下图 8.19 所示。

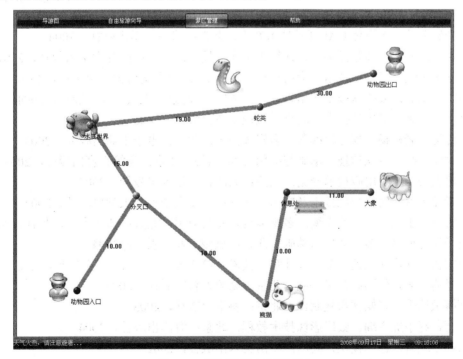

图 8.19 景区建设方案测试图

参考文献

[1] 严蔚敏，吴伟民编著．数据结构（C 语言版）．北京：清华大学出版社，1997

[2] Adam Drozdek 著．陈曙晖译．数据结构与算法（C++版）第二版．北京：清华大学出版社，2003

[3] 李建学，李光元，吴春芳编著．数据结构课程设计案例精编——用 C/C++描述．北京：清华大学出版社，2007

[4] 侯识忠编著．数据结构算法 Visual C++6.0 程序集．北京：中国水利水电出版社，2005

[5] 汪杰编著．数据结构经典算法实现与习题解答．北京：人民邮电出版社，2004

[6] 李业丽，郑良斌编著．数据结构（C）实验教程．北京：北京理工大学出版社，2005

[7] 张红霞编著．数据结构教程与实训．北京：北京理工大学出版社，2006

[8] 苏仕华编著．数据结构课程设计．北京：机械工业出版社，2005

[9] 付百文编著．数据结构实训教程．北京：科学出版社，2005

[10] 王红梅编著．数据结构（C++版）学习辅导与实验指导．北京：清华大学出版社，2005

[11] 阮宏编著．数据结构实践指导教程（C 语言版）．武汉：华中科技大学出版社，2004

[12] 张红霞，白桂梅编著．数据结构与实训．北京：电子工业出版社，2008

[13]（美）Larry R. Nyhoff 著．陈佩佩，李国东，黄达明译．C++数据结构导引．北京：清华大学出版社，2005

[14] 刘光然编著．多媒体 CAI 开发技术教程．北京：清华大学出版社，2004

[15] 王岚，刘光然，曲建民编著．计算机多媒体技术．天津：南开大学出版社，2001

[16] 王松年主编．多媒体技术与应用教程（第二版）．上海交通大学出版社，2000

[17] 赵子江编著．多媒体技术应用教程（第 4 版）．北京：机械工业出版社，2004

[18] 陈文华编著．多媒体技术．北京：机械工业出版社，2003

[19] 吴玲达，老松杨，魏迎梅编著．多媒体技术．北京：电子工业出版社，2003

[20] 鄂大伟主编．多媒体技术基础与应用（第二版）．北京：高等教育出版社，2003

[21] 尹建国著．电子出版技术导论．天津：天津科学技术出版社，2007

[22] 詹青龙主编．网络视频技术及应用．西安：西安电子科技大学出版社，2004

[23] 李大友，王岚，刘光然主编．多媒体技术及其应用．北京：清华大学出版社，2001

[24] 赵子江等编著．多媒体技术基础．北京：机械工业出版社，2004

[25] 鲁宏伟，汪厚祥主编．多媒体计算机技术（第 2 版）．北京：电子工业出版社，2004

[26] 游泽清．多媒体画面艺术基础．北京：高等教育出版社，2003

[27] 刘毓敏编著．多媒体影视制作．北京：科学出版社，2003

[28] 罗智，赵小虎主编．数码影像技术教程．北京：海洋出版社，2004

[29] 朱和平编著．艺术概论．湖南：湖南美术出版社，2002

[30] 李南著．影视声音艺术．北京：中国广播电视出版社，2001

[31] 杨改学. 艺术基础（美术）. 北京：高等教育出版社，1995
[32] 方其桂主编. Flash MX 课件制作方法与技巧. 北京：人民邮电出版社，2003
[33] 杨帆，刘鹏. 中文版 Flash MX 标准教程. 北京：中国电力出版社，2003
[34] 王定，王祥仲.中文版 Flash MX 实用培训教程.北京：清华大学出版社，2003
[35] 刘光然. 通用型 CAI 课件脚本的编写格式. 见：邱玉辉等主编. 计算机与教育——迎接 21 世纪教育信息化的挑战. 重庆：西南师范大学出版社，1999
[36] http://www.emmedu.net/（益美教育科研网）
[37] http://www.ebubu.cn:8010/（真源多媒体教育资源库）
[38] http://content.edu.tw/（台湾学习加油站）
[39] 全国计算机技术与软件专业技术资格（水平）考试办公室编. 软件设计师考试大纲. 北京：清华大学出版社，2004
[40] 全国计算机信息高新技术考试教材编写委员会编. 图形图像处理（Photoshop7.0 平台）职业技能培训教程. 北京：北京希望电子出版社，2004
[41] 全国计算机信息高新技术考试教材编写委员会编. 多媒体软件制作（Authorware 6.5）职业技能培训教程. 北京：北京希望电子出版社，2003
[42] 全国计算机信息高新技术考试教材编写委员会编. 网页制作（Flash MX 2004 平台）职业技能培训教程. 北京：北京希望电子出版社，2004